JOURNAL OF
MACHINE TO MACHINE
COMMUNICATIONS

Volume 1, No. 1 (January 2014)

JOURNAL OF MACHINE TO MACHINE COMMUNICATIONS

Editor-in-Chief
Johnson I Agbinya, La Trobe University, Australia

Aims and Scope

Currently significant research publications in machine to machine communications, Internet of things, distributed and ubiquitous computing are spread across many disciplines and journals with each one focusing on a narrow aspect of the field of Machine to Machine Intelligence. Consequently finding a single point of focus which makes it easy for researchers and in particular early career researchers to latch onto new publications in this field in one place has been difficult. The objective of Journal of Machine to Machine Communications is to provide such a focus for state of the art research findings and development is machine to machine systems. The Journal will receive contributions which deal with all aspects of machine to machine (M2M) communications, Internet of things (IoT), distributed and ubiquitous computing, communications and sensing. Towards that end the team of international editors and advisory board is formed to include individuals with extensive expertise in at least one aspect of the Journal spread. The Journal is published by the Association of International Scientists (AIS).

JOURNAL OF MACHINE TO MACHINE
COMMUNICATIONS

Volume 1, No. 1 (January 2014)

Published, sold and distributed by:
River Publishers
Niels Jernes Vej 10
9220 Aalborg Ø
Denmark

Tel.: +45369953197
www.riverpublishers.com

Journal of Machine to Machine Communications is published three times a year.
Publication programme, 2014: Volume 1 (3 issues)

ISSN: 2246–137X (Print Version)
ISBN:978-87-93102-68-2 (this issue)

Editorial foreword

The emerging new paradigm of Machine-to-machine (M2M) communications (Internet of Things) provides the basis for machines and devices to communicate with each other seamlessly worldwide. Machines are used in unimaginable areas worldwide and more and more of them are being hooked onto the Internet thereby creating a colossal wide area communication network at the global scale. Components of M2M communication systems are found in defence, agriculture, sensor networks, smart grid, smart cities, home automation and smart homes, health and transportation.

As M2M communication grows more and more new technologies and applications which underpin their performance are being created. The Journal of Machine to Machine Communications has been created to report on the state of art technologies related to M2M.

In this historical first issue we present five papers which report on both the applications and technologies that support M2M. The papers report on new developments in 'border-post cargo clearance' system, power sources for M2M and applications on time series and processing of ultrasonic images for M2M.

E. Bhero and A Hoffman report on the application of RFID systems at the air cargo terminals as a means of tracking and optimising the distribution of cargo. RFID systems are going to be used more and more in M2M networks and hence how to use them becomes an essential foundation for future development of M2M. The example provided by these authors has set the pace for more similar and varied applications of RFIDs.

How to power sensor networks in open fields and in remote areas is an interesting and essential technology which will help with rapid deployment and acceptance of M2M. The two papers by A.O. Anele, Y Hammam, A Yasser and K Djouani describe how to improve the performance of wireless power transfer systems which can be used for in-field powering of sensors and sensor networks.

One of the major areas which M2M will play a large role is in health networks. Broadband health networks require biological image processing and compression. Compressed x-ray images will be transmitted through broadband M2M networks. The paper by R. Gupta, I. Elamvazuthi, I. Faye, P. Vasant and J. George presents an example of how ultrasound images may be processed for transmission and decision support in health networks. Ultrasound images will assist "doctors in the diagnosis of medical health of patient in obstetrics, cardiology, gynaecology, musculoskeletal, urology and others". The paper is an attempt by the authors to remove speckle noise which affects ultrasound "image resolution and contrast'. Speckle noise has adverse impact on the ability of doctors in diagnosis.

In general the future of the telecommunication networks will be dominated by machine to machine technologies. In the near future operators will offer services in M2M which will greatly impact how we live, communicate and associate with other human beings and objects. It is our interest to present the emerging developments in the area as soon as they are made available to us.

Optimizing Border-Post Cargo Clearance with Auto-ID Systems

Ernest Bhero and Alwyn Hoffman

The Discipline of Electrical, Electronic and Computer Engineering, School of Engineering, University of KwaZulu Natal, Durban, South Africa,
bhero@ukzn.ac.za
The School of Electrical, Electronic and Computer Engineering, Faculty of Engineering, Northwest University, Potchefstroom, South Africa
alwyn.hoffman@nwu.ac.za

Received 24 August 2013; Accepted 18 December 2013
Publication 23 January 2014

Abstract

Stakeholders in cross-border logistics and trade corridors have always been concerned about ways of improving operational efficiency. Cargo owners and cargo forwarders have been particularly concerned with long delays in the processing and clearing of cargo at border-posts. It is reasonably suspected that the delays are due to a combination of a lack of optimum systems' configurations and the inefficient human-dependent operations, which makes the operations prone to corruption. This paper presents the findings of a study, which is being conducted to determine the sources of the inefficiencies and then suggest possible solution based, largely, on RFID technology. The procedure and preliminary findings are presented.

Keywords: Road Freight Transport, Cross-border, Customs Processes, GPS Tracking, RFID, OSBP, Choke Monitoring.

1 Introduction

Over the years, there has been a steady increase in levels of international freight movement, thanks to globalization. Although the bulk of the freight is ocean

Journal of Machine to Machine Communications, Vol. 1, 1–14.
doi: 10.13052/jmmc2246-137X.111

bound, road transport remains an important link in multi-modal freight supply chains [11]. On the African continent this dependence on road transport is even more prominent due to the absence or bad state of railway lines. In Africa, many countries are landlocked, which means a significant portion of road freight must travel along multinational corridors [11]. While many economic regions are gradually doing away with the charging of customs duties at border posts [12], the opposite is true in Africa, where the majority of countries are still dependent on customs duties as their primary source of state income [7]. Stringent controls are, therefore, applied at most border posts to ensure that freight does not leak into or from a country before the required duties are paid.

The effective management of road transport is complicated by the involvement of a wide spectrum of independent stakeholders. The cross-border movement of a freight consignment includes at least the cargo owner (or consignor), the transport company, potentially a transport broker, a freight forwarder, a clearing agent, roads agencies, customs authorities on each side of the border, and a customer (or consignee). While the commercial players have objectives that are well aligned and aimed at operational efficiencies, the agencies responsible for protecting roads infrastructure and collecting customs duties are more concerned about the effectiveness of control measures aimed at the prevention of illegal practices.

Current cross-border freight management systems are characterized by a lack of transparency from the perspective of the consignor and consignee and by little coordination between the actions of different role-players. The lack of visibility of operations at ground level results in long average delays at border posts accompanied by many corrupt practices - often these two go hand in hand [1, 8]. While lacking integration between the systems operated by different stakeholders is partly to blame, deliberate manipulation of the process by human operators (who know that their actions are difficult to police) also plays a major role.

This paper will focus on cross-border operations of typical border-posts in sub-Sahara Africa. Firstly, the paper analyzes practical studies performed at typical border-posts to measure the average delays experienced by freight consignments [9, 10].

Secondly, the paper proposes an improved cross-border management concept, based on the use of information and communication technologies and on the integration between the systems operated by private and public sector stakeholders.

Thirdly, the paper presents possible improvements on border-posts operations based on related studies.

2 Literature Review

In this section, we will consider, firstly, the empirical work done to evaluate the current state of affairs at selected border-posts and the related theoretical studies of possible improved scenarios.

2.1 Border-Post Operations Field-Studies

Extensive work has been done to determine work flows at border-posts and to determine the major contributing factors for the undue delays experienced at border-posts [11]. In this presentation, we will focus on the Chirundu border-post because Chirundu is now a One Stop Border-Post (OSBP). OSBPs have been suggested as alternative configurations to traditional border-posts; the rationale for OSBP was that it would improve operational efficiencies of border-posts. Table 1 summarizes the field-work done at Chirundu border-post, [9, 10].

The table shows time-delays for various border activities before and after the border was turned into an OSBP. The summary of the field work indicates that:

- Customs, Document Processing and Agent times have actually doubled.
- Driver Idle Time has increased by about 27%.
- Not apparent from Table 1, 50% of all vehicles crossed the border within 24 hours.

Table 1

TMSA Choke Monitoring at Chirundu

Comparative Category	Current OSBP Data	Comparative Pre-OSBP Data
Summary of Bottlenecks in Hours – North bound		
Agent	12:31	6:00
Customs	45:27	21:00
Driver Idle Time	14:06	11:00
Immigration	0:31	0.00
Document Processing	57:58	27:00
Inspection Time	1:14	1:00

These observations among others indicate that the objective of establishing an OSBP was not entirely achieved; in fact the objective seems to have been completely eroded and reversed. However, the fact that 50% of the vehicles crossed the border within 24 hours comes as consolation.

The report of the work, [9, 10], confirmed the conviction that, the apparent failure of this OSBP is due mainly to negative human conduct. This observation further strengthens the belief that, the possible solution lies in automating most of the operations and closely monitoring human conduct in a way that fosters accountability.

2.2 Trade Facilitation Objectives of Sub-Sahara Regional Bodies

In a COMESA–SADC meeting held in Gaborone, Botswana on the 3rd and 4th of February 2011, the objectives of the Regional Economic Communities (REC) as far as trade facilitation was clearly spelt-out, [14]. The mission objectives read:

> "The main objective of the meeting was to finalize and adopt a Joint Strategy and Work Plan for the development and implementation of a joint SADCCOMESA CUSTOMS Transit Management Information System and to discuss and possibly come up with a road map on the simplification and harmonization of a Transit Management System in the regions."

The findings and the report of that meeting indicated that, there is political will not only at Governmental levels but at regional levels too to facilitate harmonious trade between member countries. The meeting also brought out the imperative nature of the need to optimize trade processes in the region. From the report, it is apparent that, the RECs aim, mostly, at harmonizing information interchange at customs level and not at improving operational efficiencies at other levels such cross-border operations. Thus, there is need to compliment the REC's effort in this regard.

2.3 Related Studies on Trade Corridors' Operations Optimization

The most notable study in a related area was done by Hsu *et al* [2]. This study looked at import cargo processing in an air cargo terminal. It then constructed a customs clearance-network based on cargo, information and human flows. The flow network lead to the sub-division of the network into several operational units and a customs clearing team would work on separate units. The study then formulated a mathematical model for describing the customs clearance process delay and how it affects the delay of cargo arriving later. The performance at the cargo terminal was then assessed and analyzed on the basis of this model. RFID system was then introduced in relevant operational units such as replacing some labour operations, identification function and position locating functions. The decrease in cargo owner's inventory cost and terminal operators' labour cost were evaluated to measure the gains of reducing delays due to the use of RFID system at the air cargo terminal. The analysis, indicated performance of about 63%. However, the researchers indicated a decrease in percentage when the volume of cargo handled per given time increased beyond a certain limit. This decrease in performance is attributed to the limit in the number of work teams working on customs clearance.

Although this work, [2], was carried out at an air cargo terminal, it bears close resemblance with cross-border operations. We will present a detailed comparison in the subsequent sections and try and draw insight into cross-border operations.

3 Analysis of Border Post Operations

In this section, we will analyse the typical process flow diagram at a border-post, analyse the typical border-post layout and then identify possible process flow bottle necks.

3.1 Typical OSBP Border Post Process Flow Diagram

Figure 1 below shows the process flow diagram for OSBP (for example, Chirundu).

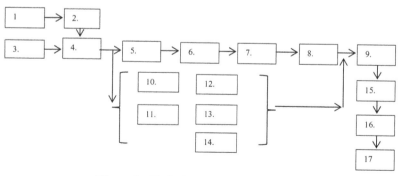

Figure 1   Typical Border-post Work Flow

Table 2

1. Lodging of manifest for preclearance.	8. Customs process clearance.	15. Customs release consignment (docs to agent).
2. Pre-clearance and pre-payment.	9. Payment of customs duties and other payments.	16. Agent hands documents to driver.
3. Travel from origin to border post.	10. Driver goes through immigration procedures.	17. Consignment leaves the border / cross to other side.
4. Consignment arrives at border post.	11. Vehicle weighed at weigh-bridge.	
5. Driver submits documents to agent.	12. Vehicle scanned at scanning shed.	
6. Agent prepares documents for submission to customs.	13. Physical inspection of consignment.	
7. Agent submits documents to customs.	14. Inspections by other border agencies.	

The activities in the flow chart are summarized in Table 2, [10]:

From Table 1 we note that, activities 6, 7, 8, 9 and 15 contribute immensely to the delays observed at the border-posts. As mentioned earlier, the delays observed at the border-posts are due to a combination of factors; human and system configurations. Thus, the solution sort is multifaceted; finding and suggesting improved system configurations and minimizing human involvement and closely monitoring human activities by use of auto-ID systems lead by RFID.

3.2 Typical Border Post Layout

Figure 2 below shows typical border-post layout.

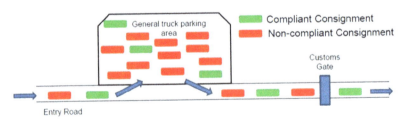

Figure 2   Typical Current Border Post Lay-out

In this layout, compliant consignment and non-compliant consignment queue into and out of the border using a single lane. There is parking yard for both consignments. In this configuration, it often happens that, a compliant consignment is blocked by a non-compliant truck (consignment). The end result is that, the compliant consignment gets unduly delayed just like the non-compliant consignment. At the end of the day, the benefits for being compliant are diminished and therefore would save no purpose in being compliant.

4 Proposed Solution and Simulated Results

We begin this section by presenting a proposal for border-post layout re-configuration and why that change is presumed beneficial. We then propose an RFID-based system to automate certain border-post operations with a presumed improvement in operational efficiency at border-posts. We conclude this section by presenting a comparative study of border-post operations as compared to the work of Hsu *et al* [2] and thereby presenting the simulated results.

4.1 Proposed Layout Re-configuration

Figure 3 shows a modified border-post layout. The rationale for this layout is that, compliant consignment should not be unduly delayed. On entering the customs/immigration yard, the road is split into two, the "green lane" for compliant cargo and "red lane" for non-compliant or suspicious cargo. At the entry gate, there will be RFID readers, which automatically read tags from entering trucks and determine the risk status of the cargo (whether compliant or not). The Auto-ID system will then open appropriate boom gate into the green lane or the red lane. Thus, the compliant cargo will move freely and fast through the customs/immigration area. The non-compliant cargo will have to

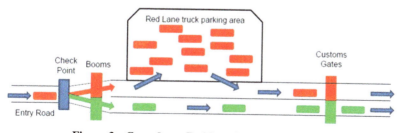

Figure 3   Green Lane /Red Lane Border Post Layout

go to the parking yard for inspection. It should be noted that, green lane layout will work better with pre-clearing and pre-payment as indicated in Figure 1 above. What else can be done to further enhance the green lane concept?

4.2 Proposed Pro-RFID System Assisted Operations

4.2.1 RFID technology

RFID means Radio Frequency Identification. It is an emerging and independent interdisciplinary field. It combines technologies such as HF technology and EMC, data protection and cryptography, semiconductor technology, telecommunications and other related areas. RFID is widely used, if not exclusively used, in automatic identification procedures. The Auto-ID procedures can provide information about people, animals, goods and products in transit. RFID system comprises of RFID readers and a variety of transponders or tags. The tags can either be passive or active tags. Tags are also categorized in terms of their frequency of operation. The RFID readers establish a link between back-end system (databases and networks) and the front-end system, which is made up of tags and other related auto-id procedures. The RFID reader wirelessly communicate and extract information/data from tags and relay this information for processing in the back-end system or in a standalone system, the reader can process the data/information and carryout required tasks.

4.2.2 The proposed RFID-based system

Auto-identification systems have existed since the 1960s when the barcode system was first introduced. Today we have a wide range of auto-identification systems in the market and RFID has been growing fast as an auto-identification system of choice in wide-ranging applications. But, according to, [15], *technology adoption is not always about choosing the dominant design but about how to future-proof an auto-ID implementation.* Thus, although the researchers suggest a predominantly RFID based auto-id system in this work, the researchers are mindful of the need to incorporate other auto-id systems and indeed that arm of work is being researched on too.

From our analysis of Table 1 and Figure 1, it is apparent that the customs processes, document processing, which is done by the customs officials and the Agents contribute the longest delays. Thus, we propose the use of electronic documents or to have documents that are embedded with an RFID tag. The tag will contain information like what is normally found on the manifest document and declaration or consignment data. This allows Electronic Data Interchange (EDI) between the cargo owners or their Agents and the customs authorities.

Further, the cargo itself will be tagged and this allows the RFID readers at the entrance to read the tags and identify the cargo as it arrives. The cargo identity is then feed to the control centre, which then initiates immediate processing of the consignment. The risk engines of the customs will then determine the risk level of the consignment and alert the inspection team whether to physically inspect a particular cargo or not. All the information interchange is done electronically thereby improving the rate of document processing. The human labour removed from document processing can be deployed in the inspection bays and monitoring the entrance and exit points. Implementing the system as just described will substantially remove human involvement in the document processing and this is expected to improve the performance at the border-posts.

4.2.3 Comparative study with the research work in [2]

Performance of envisaged system was assessed solely on the basis of closely related work such as [2]. The related work in [2] was done at an air terminal (airport) in Taiwan; yet in this study it is the border-posts which are the main focus. The following subsections discuss the differences and similarities between the two studies. The comparisons enable us to predict the possible performance improvement for the border-posts.

- *Differences*:
    i. Air terminals handle mostly high value and perishable goods.
    ii. At air terminals, cargo is removed from the aeroplane and temporarily stored by customs during processing and inspection.
    iii. At air terminals, there are no trucks carrying cargo.
    iv. Air terminals handle less cargo per given time compared to border-posts.
- *Similarities*:
    i. In both systems we have; information flow, human flow and cargo flow paths.
    ii. Both systems have document processing by customs and/or Agents or customs brokers.
- *Simulated Results*:

In the analysis by Hsu *et al* [2], the following equation was formulated to describe time taken by c^{th} cargo to go through various clearance processes.

$$t_{u,r}^c = Max\left\{A_{u,r}^{c-1} + t_{u,r}^{c-1} - A_{u,r}^c - S_{c-1,c}, 0\right\} + I_{u,r}^c \quad \forall u, r, c \quad (1)$$

Where:

- $t_{u,r}^c$ is the total time taken by c^{th} cargo to complete activities of operation at stage u handled by working team r.
- $I_{u,r}^c$ is the cargo handling and waiting time.
- $A_{u,r}^c$ is the accumulative time taken by cargo for completing all upstream activities at parts 1 to u-1. Note, $A_{1,r}^c = 0$.
- $S_{c-1,c}$ is the time headway between $(c-1)^{\text{th}}$ and c^{th} cargos.

According to the study, [2], the cargo at the Taiwan air cargo terminal goes through a total of six stages of clearing operations. With reference to Figure 1, we see that there is a possible maximum of four stages of clearing operations; if the 'other inspections' (block diagram 14) and weighing (block diagram 11) are assumed to be negligible there will only be two stages. Thus, for compliant cargo, the delay bull-whip effect will be minimal for border-posts compared to air terminals. Using time delays in Table 1 together with process flow diagram in Figure 1, we have Figure 4, which shows the possible resultant improvement on processing time when an auto-id system lead by RFID is implemented.

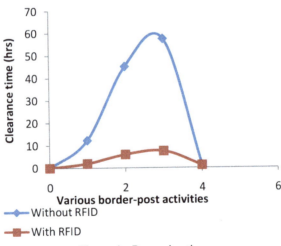

Figure 4 Processing time

5 Discussions and Conclusions

The expected performance improvement depicted in Figure 4 has assumed an ideal auto-id system. It has also assumed no negative human interference. In reality individuals can delay capturing or scanning documents. Therefore, there is need to bring a system with a human activity monitoring functionality to the proposed system. Also, the system will be more complete if it can monitor movement of cargo throughout the transit period of cargo from source to destination. If the tracking data is shared or linked to customs' cargo risk engines, then it becomes easier to separate compliant from non-compliant cargo. This would further enhance the usefulness of the system with the overall expected improvement in trade facilitation. However, a more detailed modelling of every stage is required in order to come up with a system close to realistic settings. Further work being done include; investigating impact of current ICT paradigms on the implementation of the envisaged system and what levels integration of various systems used by different stakeholders is realistically possible.

References

[1] http://ieg.worldbankgroup.org/content/dam/ieg/gac/gac_eval.pdf World Bank Country-Level Engagement on Governance and Anticorruption. Accessed 9th of April (2013).

[2] C. I. Hsu, H. H. Shih and W. C. Wang, "Applying RFID to reduce delay in import cargo customs clearance process". Computers & Industrial Engineering Volume **57,** (2009).

[3] J. Siror, S. Huanye, W. Dong and W. Jie, "Application of RFID technology to curb diversion of transit goods in Kenya". Fifth International Conference on INC, IMS and IDC (2009).

[4] J. Siror, S. Huanye and W. Dong, "Evaluation of RFID Based Tracking Systems for Securing Transit Goods in East Africa". Sixth International Conference on Digital Content, Multimedia Technology and its Applications (2010).

[5] J. Siror, S. Huanye, W. Dong, L. Guangun and P. Kaifang, "Impact of RFID technology on tracking of export goods in Kenya". Journal of convergence information technology Volume **5** number nine. (2010).

[6] B. Li and W. F. Li, 2010, "Modeling and Simulation of Container Terminal Logistics Systems Using Harvard Architecture and Agent-Based Computing". Proceedings of the 2010 Winter Simulation Conference.

[7] A. J. Hoffman, "The use of technology for trade corridor management in Africa", NEPAD Transport Summit, Sandton, South Africa, October (2010).

[8] "Bribery as a non-tariff barrier to trade – a case study of East Africa trade corridors", Transparency International-Kenya (2012).

[9] M. Fitzmaurice, "Optimization plan for freight movements at key commercial border crossings", Report compiled by Transport Logistics Consultants, August (2009).

[10] M. Fitzmaurice, "TLC Report on Beitbridge, Chirundu, Kasumbalesa and Nakonde", November (2012).

[11] S. Mpata, "Evaluation of the COMESA/SADC transit management systems – Final Report", Lilongwe, Malawi, September (2011).

[12] L. Norov and D. Akbarov, "Customs – time for a change", Problems of Economic Transition, vol. **52**, no. 2, pp. 47–60, June (2009).

[13] http://www.afdb.org/fileadmin/uploads/afdb/Documents/Knowledge/20 09%20AEC-%20Towards%20an%20EAC%20COMESA%20and%20S ADC%20Free%20Trade%20Area%20Issues%20and%20Challenges.pd f Evaluation of the COMESA/SADC Transit Management Systems: September, 2011. Accessed 21/04/2013.

[14] http://www.trademarksa.org/sites/default/files/publications/Mission%20 Report%20%7C%20COMESA-ADC%20Cooperation%20on%20Custo ms%20Transit%20Management%20Information%20System%20%7C% 20February%202011.pdf Mission Report on COMESA-SADC Cooperation on Customs Transit Management Information System. Accessed 21/04/2013.

[15] http://ieeexplore.ieee.org/stamp/stamp.jsp?tp=&arnumber=4037179 The Hybridization of Automatic Identification. Accessed 02-01-2012

[16] S. R. Hong, S. T. Kim, C. O. Kim, "Neural network controller with on-line inventory feedback data in RFID-enabled supply chain", International Journal of Production Research, volume **48,** November (2009).

[17] A. Ustundag, "Evaluating RFID investment on a supply chain using tagging cost sharing factor", International Journal of Production Research, volume **48**. November (2009).

[18] S. J. Wang, C. T. Huang, W. L. Wang, Y. H. Chen, "Incorporating ARIMA forecasting and service-level based replacement in RFID-enabled supply chain", International Journal of Production Research, volume **48**, November (2009).

[19] N. Langer, C. Forman, S. Kekre, A. Scheller-Wof, "Assessing the Impact of RFID on Return Centre Logistics", Interfaces Vol-**37**, December (2007).

Biographies

Ernest Bhero is a lecturer in the Discipline of Electrical and Electronic Engineering at the University of KwaZulu Natal, Durban, South Africa. He holds B. Eng. (Electronics), MPhil (Electronics) and a PhD candidate with North-West University. As an Engineer, he designed and developed three electronic products that were commercialised in Zimbabwe. The areas of interest and research include; Computer Architecture and Organization, Embedded Systems and RFID Systems and their application in improving efficiency in trade corridors' operations.

Alwyn Hoffman was born in Bloemfontein, South Africa in 1962. He received the B. Eng. (Electronics) degree in 1985, the M. Eng. (Electronics) in 1987, the Ph. D. (Electronics) in 1991 and the MBA degree in 1996, all from the University of Pretoria. After working in the South African high technology electronics industry during the period 1985 to 1994, he joined Northwest University in October 1994 as Director of the School of Electrical and Electronic Engineering for a period of 6 years. Between 2001 and 2008 he again spent time in the high-technology industry, working for Inala Technology Investments during 2001 in the field of intellectual property development and for IPICO Inc., a company listed on the Toronto Stock Exchange, during the period 2002 till 2008 in the field of automated identification. In January 2009 he rejoined Northwest University's Faculty of Engineering, and is still involved as consultant in the local and global electronics and communication industries. His research field includes wireless communications and artificial intelligence.

Effects of Coil Misalignments on the Magnetic Field and Magnetic Force Components between Circular Filaments

Anele O. Amos[1,2], Hamam Yskandar[1,2,3], Alayli Yasser[2]
and Djouani Karim[1,4]

[1]Dept. of Electrical Engineering, Tshwane University of Technology, Pretoria, South Africa
[2]LISV Laboratory, UVSQ, Paris, France
[3]ESIEE Paris Est University, Paris, France
[4]LISSI Laboratory Paris Est University, Paris, France
anelea@tut.ac.za, hamama@tut.ac.za, yasser.alayli@lisv.uvsq.fr,
djouanik@tut.ac.za

Received: 24 August 2013; Accepted 18 December 2013
Publication 23 January 2014

Abstract

Wireless transfer of electrical energy between air-cored circular coils can be achieved using contactless inductive power transfer (CIPT) system. Notwithstanding, the aim of designing CIPT systems is not always realized because of coil misalignments between the primary and the secondary coils. Based on this information, the author of this paper analyses the effects of lateral and angular misalignments on the magnetic field and the magnetic force between filamentary circular coils which are arbitrarily positioned in space. This investigation is achieved based on advanced and relevant models formulated in the literature. Detailed results obtained using SCILAB application software is given.

Keywords: Circular Filaments, Coil Misalignments, Magnetic Fields, Magnetic Forces.

Journal of Machine to Machine Communications, Vol. 1, 15–34.
doi: 10.13052/jmmc2246-137X.112

1 Introduction

Contactless inductive power transfer (CIPT) systems are a modern technology used for transferring electrical energy over a relatively large air-gap via high frequency magnetic fields. The wireless transfer of electrical energy to the on-board battery storage system of electric vehicles (EVs) is obtained via CIPT transformer which consists of air-cored coils. Air-cored CIPT systems are preferred to an iron-cored type due to the design objectives of controllability. In addition, high power transfer capability may be achieved since the compensation networks of the primary and the secondary coils are not considered independently [1]. However, the purpose of CIPT systems is not always attainable due to lateral and angular misalignment between the primary and the secondary coils.

A number of authors [2–8] have dealt with the computation of magnetic force between coaxial circular coils using analytical methods. However, the current focus is on the computation of magnetic force between circular coils with lateral and angular misalignments [9–16]. With the use of finite element and boundary element methods, magnetic fields and forces can be accurately and rapidly computed [17, 18]. Notwithstanding, the authors in [19] argue that the computation of these important physical quantities can be solved using semi-analytical methods since they considerably reduce the computational time and the enormous mathematical procedures. Also, it is argued that the formulated models obtained using the Lorentz law represent the simplification of the models obtained by the Biot-Savart law and the mutual inductance approach [20]. Finally, it is concluded that the formulated model is easy to understand, numerically suitable and easily applicable for engineers and physicists.

In this paper, the authors investigate the effects of arbitrary lateral and angular misalignment on the magnetic field components which exist when current flows through the primary coil and the magnetic force components exerted on the current carrying conductor. The computations of these important physical quantities between circular filaments arbitrarily positioned in space are achieved based on the advanced and relevant models given in [19]. To achieve this task, this paper is organized as follows. Section 2 presents the advanced and relevant models formulated in the literature. In section 3, results obtained using SCILAB are given. Section 4 discusses the results and section 5 concludes the paper.

2 Formulated Advanced and Relevant Models

This section presents the 3-D space positions of circular filaments with coil misalignments, geometric configurations and common notations for circular filaments and the advanced and relevant models for computing the magnetic field and magnetic force components between circular filaments.

2.1 3-D Space Positions for Circular Filaments with Coil Misalignments [19]

Shown in Figure 1 are the circular filaments with coil misalignments. The secondary side (i.e., the smaller circle) which is placed in an inclined plane $x'Cy'$ is laterally and angularly misaligned whereas the centre of the primary coil is placed at the plane xOy with the axis of z along the axis of the larger circle. The general equation for the smaller circle is given as

$$\lambda = ax + by + cz + D = 0 \tag{1}$$

The centre of the secondary coil is defined in the plane λ, C (x_C, y_C, z_C) and the coordinates of the point D_S which are given as

$$D_1 \left[x_C - \frac{abR_S}{Ll}, \quad y_C + \left(a^2 + c^2\right)\frac{R_S}{Ll}, \quad z_C - \frac{bcR_S}{Ll} \right]$$

or

$$D_2 \left[x_C + \frac{abR_S}{Ll}, \quad y_C - \left(a^2 + c^2\right)\frac{R_S}{Ll}, \quad z_C + \frac{bcR_S}{Ll} \right]$$

where

$$L = \left(a^2 + b^2 + c^2\right)^{0.5}, l = \left(a^2 + c^2\right)^{0.5} \text{[10]}.$$

The following are the 3-D space positions required for the computation of the magnetic field and the magnetic force between circular filaments with lateral and angular misalignments [19]:

- The primary coil of radius R_P is placed in the plane xOy $(z = 0)$ with the centre at O $(0, 0, 0)$. An arbitrary point $B_P(x_P, y_P, z_P)$ of this coil has parametric coordinates which are given as

$$
\begin{aligned}
x_P &= R_P \cos t \\
y_P &= R_P \sin t \qquad\qquad t \in (0, 2\pi) \\
z_P &= 0
\end{aligned}
\tag{2}
$$

- The differential element of the primary coil is given by

$$d\vec{l}_P = R_P \left(-\vec{i}\sin{(t)} + \vec{j}\cos{(t)}\right) dt \qquad t \in (0; \ 2\pi) \qquad (3)$$

- The unit vector N (i.e., the unit vector of the axis z') at the point C (centre of the secondary coil) laying in the plane λ, is defined by

$$\vec{N} = \{n_x, \ n_y, \ n_z\} = \left\{\frac{a}{|\vec{n}|}, \ \frac{b}{|\vec{n}|}, \ \frac{c}{|\vec{n}|}\right\}$$
$$|\vec{n}| = \left(a^2 + b^2 + c^2\right)^{0.5} = L \qquad (4)$$

- The unit vector between two points C and D_S placed in the plane λ is

$$\vec{u} = \{u_x, u_y, u_z\} = \left\{-\frac{ab}{lL}, \ \frac{l}{L}, \ -\frac{bc}{lL}\right\}$$
$$L = \left(a^2 + b^2 + c^2\right)^{0.5}, \quad l = \left(a^2 + c^2\right)^{0.5} \qquad (5)$$

- The unit vector v is defined as the cross product of the unit vectors N and u as follows

$$v = \vec{N} \times \vec{u} = \{v_x, v_y, v_z\} = \left\{-\frac{c}{l}, \ 0, \ \frac{a}{l}\right\}$$
$$l = \left(a^2 + c^2\right)^{0.5} \qquad (6)$$

- An arbitrary point E_S (x_S, y_S, z_S) of the secondary coil has parametric coordinates which is a well-known parametric equation of circle in 3-D space and is given as

$$\begin{aligned} x_S &= x_C + R_S u_x \cos\phi + R_S v_x \sin\phi \\ y_S &= y_C + R_S u_y \cos\phi + R_S v_y \sin\phi \quad \phi \in (0, \ 2\pi) \\ z_S &= z_C + R_S u_z \cos\phi + R_S v_z \sin\phi \end{aligned} \qquad (7)$$

- The differential element of the secondary coil is also given by

$$d\vec{l}_S = R_S \left\{l_{S_x}\vec{i} + l_{S_y}\vec{j} + l_{S_z}\vec{k}\right\} d\ \phi \quad \phi \in (0, 2\pi) \qquad (8)$$

where
$$\begin{aligned} l_{Sx} &= -u_x \sin\phi + v_x \cos\phi \\ l_{Sy} &= -u_y \sin\phi + v_y \cos\phi \\ l_{Sz} &= -u_z \sin\phi + v_z \cos\phi \end{aligned}$$

2.2 Geometric Configurations and Common Notations [10]

The geometric configurations for circular filaments with air-core are given in Figures 2 – 4.

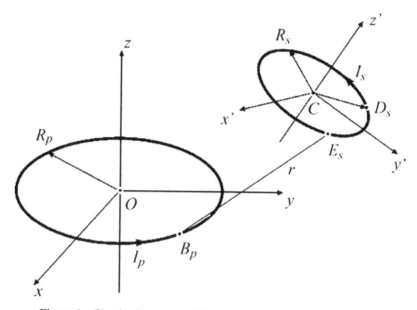

Figure 1   Circular filaments with lateral and angular misalignment [10]

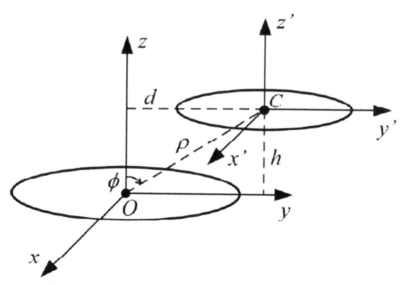

Figure 2   Circular filaments with lateral misalignment only [10]

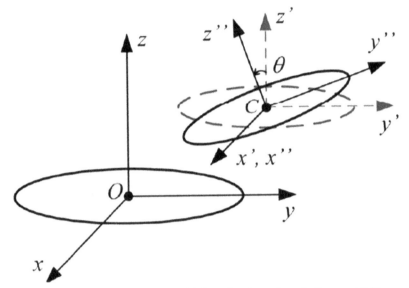

Figure 3   Circular filaments with lateral and angular misalignment [10]

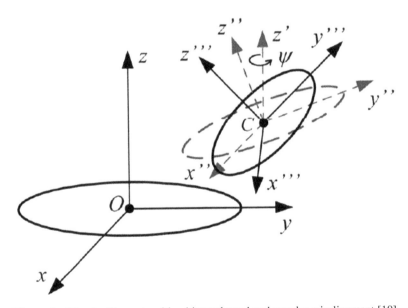

Figure 4   Circular filaments with arbitrary lateral and angular misalignment [10]

Regarding Figure 2, $\theta = 0$ and $\phi = 0$ at axes $y-z$ and $y'-z'$ coplanar. d is the horizontal distance between their centres, h is the vertical distance between their centres, ρ is the lateral misalignment (i.e., distance between their centres), ϕ is the variable rotation angle at any point of the secondary coil. Also, in Figure 3, $\phi = 0$ since axes $y - z$ and $y'' - z''$ are coplanar. θ is the angle of inclination between circular coils. Finally, regarding Figure 4, there are no coplanar axes. Therefore, to achieve the aim of this paper, Figure 4 is considered amongst other geometric configurations.

The following common notations are introduced in [10] to make easier link with [21]:

- The centre of the secondary coil must be taken at point

$$C\left(x_C = 0, y_C = d, z_C = h\right)$$

where

$$d = x_2 \sin \theta$$
$$h = x_1 - x_2 \cos \theta \qquad (9)$$
$$\rho = \sqrt{d^2 + h^2}$$
$$\cos \phi = h/\rho$$

x_1 is the vertical distance when axis z'' intersects with axis z from the origin O and x_2 is the distance between the intersection and the centre of the secondary coil

- The equivalence between Grover's latitude and longitude angles θ and ϕ and the a, b and c parameters defining the secondary coil plane is that of a spherical Cartesian system of coordinates [10, 21] which is

$$a = \sin \phi \sin \theta$$
$$b = - \cos \phi \sin \theta \qquad (10)$$
$$c = \cos \theta$$

To investigate the effects of lateral and angular misalignment between circular filaments that are arbitrarily positioned in space, the authors of this paper re-stated equations (5–8) in terms of the common notations given in equations (9–10).

2.3 Models for Computing Magnetic Field and Magnetic Force Components

The advanced and relevant mathematical models for computing the magnetic field and the magnetic force components between circular filaments with lateral and angular misalignment are given in [19] as follows:

- The final form of the magnetic field in an arbitrary point $D_S(x_S, y_S, z_S)$ produced by the primary coil of the radius R_P carrying current I_P is

$$B_x(x_S, y_S, z_S) = -\frac{\mu_0 I_P z_S x_S k}{8\pi\sqrt{R_P}(x_S^2+y_S^2)^{5/4}}L_0$$

$$B_y(x_S, y_S, z_S) = -\frac{\mu_0 I_P z_S y_S k}{8\pi\sqrt{R_P}(x_S^2+y_S^2)^{5/4}}L_0 \qquad (11)$$

$$B_z(x_S, y_S, z_S) = \frac{\mu_0 I_P k}{8\pi\sqrt{R_P}(x_S^2+y_S^2)^{3/4}}S_0$$

- The final form of the magnetic force components between circular filaments is also given as

$$F_x = \frac{\mu_0 I_P I_S R_S}{8\pi\sqrt{R_P}}\int_0^{2\pi} I_x d\phi$$

$$F_y = \frac{\mu_0 I_P I_S R_S}{8\pi\sqrt{R_P}}\int_0^{2\pi} I_y d\phi \qquad (12)$$

$$F_z = \frac{\mu_0 I_P I_S R_S}{8\pi\sqrt{R_P}}\int_0^{2\pi} I_z d\phi$$

where

$$I_x = \frac{k}{(x_S^2+y_S^2)^{5/4}}\left[z_S y_S l_{Sz} L_0 + \sqrt{x_S^2+y_S^2}S_0 l_{Sy}\right]$$

$$I_y = -\frac{k}{(x_S^2+y_S^2)^{5/4}}\left[z_S y_S l_{Sz} L_0 + \sqrt{x_S^2+y_S^2}S_0 l_{Sx}\right]$$

$$I_z = \frac{k}{(x_S^2+y_S^2)^{5/4}}z_S\left[x_S l_{Sy} - y_S l_{Sx}\right]L_0$$

$$k^2 = \frac{4R_P\sqrt{x_S^2+y_S^2}}{\left(R_P+\sqrt{x_S^2+y_S^2}\right)^2+z_S^2} \quad L_0 = 2K(k) - \frac{2-k^2}{1-k^2}E(k)$$

$$S_0 = 2\sqrt{x_S^2 + y_S^2}\, K\left(k\right) - \frac{2\sqrt{x_S^2 + y_S^2} - \left(R_P + \sqrt{x_S^2 + y_S^2}\right) k^2}{1 - k^2} E\left(k\right)$$

$K\left(k\right)$ and $E\left(k\right)$ are the complete integral of the first and second kind [22, 23].

3 Results Obtained

Shown in Figures 5–12 are the results obtained based on Eq. 11 and Eq. 12. These simulations are achieved using SCILAB application software [24] and the data used is shown in Table 1.

4 Discussion of Results

The effects of coil misalignments such as the lateral ρ and angular θ on the magnetic field and the magnetic force components between circular filaments arbitrarily positioned in space are shown in Figures 5–12.

The computation of the results obtained is achieved by computing equations (11) and (12) with respect to the variable rotation angle ϕ at any point of the secondary coil. Shown in Figures 5–12 are the results obtained for the magnetic field B_x, B_y and B_z components which exist when current flows through the primary coil and the magnetic force F_x, F_y and F_z components exerted on the current carrying conductor. Based on the data given in Table 1, it is clearly seen that the values of B_x, B_y, B_z and F_x, F_y, F_z components decrease and increase at certain variable rotation angles at any point of the secondary coil.

Table 1 Data used for Simulation [10]

Variable Parameters			
x_1(m)	x_2(m)	ρ(m)	θ(degree)
0.05	0.0125	0.04	30
0.10	0.0250	0.08	45
0.15	0.0375	0.14	60
0.20	0.0500	0.19	75
Constant Parameters			
R_P	0.16m		
R_S	0.10m		
μ_0	$4\pi \times 10^{-7} H/m$		

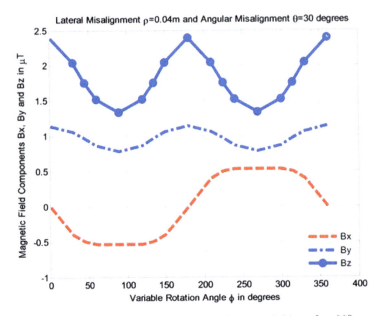

Figure 5 Magnetic field components when $\rho = 0.04m, \; \theta = 30°$

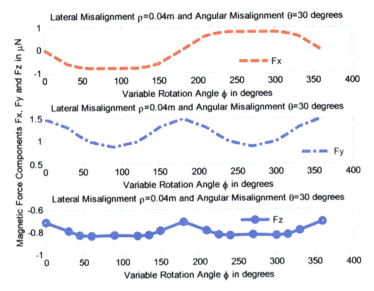

Figure 6 Magnetic force components when $\rho = 0.04m, \; \theta = 30°$

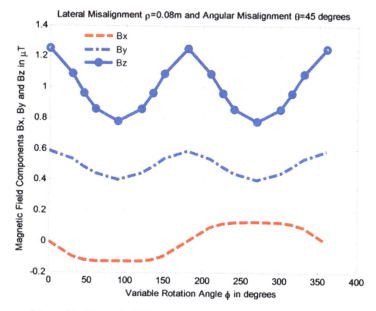

Figure 7   Magnetic field components when $\rho = 0.08m, \ \theta = 45°$

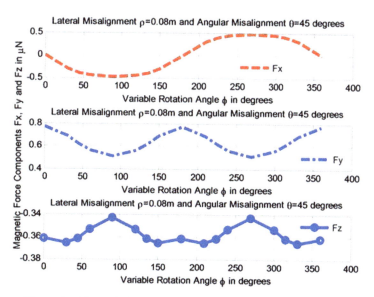

Figure 8   Magnetic force components when $\rho = 0.08m, \ \theta = 45°$

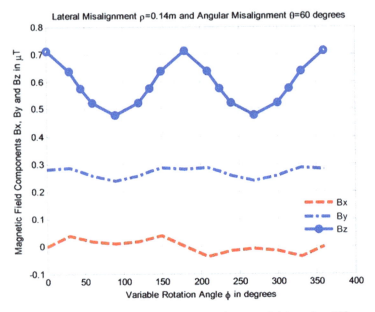

Figure 9   Magnetic field components when $\rho = 0.14m,\ \theta = 60°$

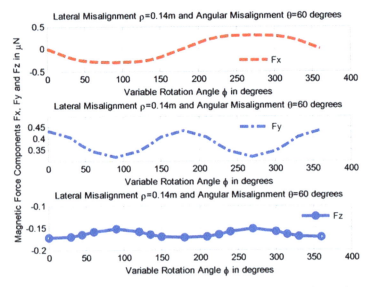

Figure 10   Magnetic force components when $\rho = 0.14m,\ \theta = 60°$

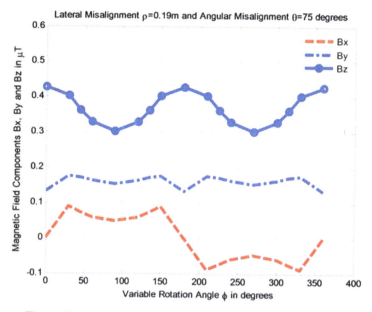

Figure 11 Magnetic field components when $\rho = 0.19m, \ \theta = 75°$

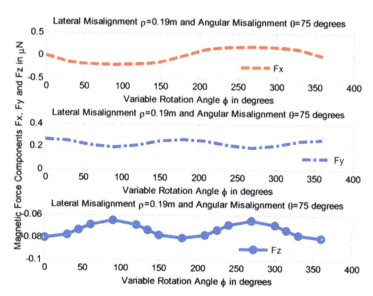

Figure 12 Magnetic force components when $\rho = 0.19m, \ \theta = 75°$

This analysis shows that in order to obtain the full potential of CIPT systems, the challenges encountered due to arbitrary lateral and angular misalignments between filamentary circular coils must be considered in the CIPT transformer model to be designed for the CIPT system.

5 Conclusion

Unlike plug-in connectors contactless inductive power transfer (CIPT) system is a modern technology which enables electrical energy to be transferred over a relatively large air-gap via high frequency magnetic fields. With the CIPT transformer, electrical energy can be transferred from the primary to the secondary air-cored circular coils. Notwithstanding, its full advantage is not always achievable due to arbitrary lateral and angular misalignments between its coils. Based on this information, this paper examined the consequences of these coil misalignments on the magnetic field and the magnetic force components between circular filaments which are arbitrarily positioned in space.

The computations are based on the advanced and relevant models formulated in the literature. The results obtained using SCILAB application software show that as the lateral and angular misalignment increase the magnetic field and magnetic force components between the circular filaments decrease and increase at certain variable rotation angles at any point of the secondary coil.

This study shows that the issues regarding arbitrary lateral and angular misalignment between circular filaments must be tackled by considering these misalignments in the model developed for the CIPT transformer.

References

[1] C. S. Wang, O. H. Stielau, and G. A. Covic, "Load models and their application in the design of loosely coupled inductive power transfer systems," in *Power System Technology, 2000. Proceedings. PowerCon 2000. International Conference on*, pp. 1053–1058 (2000).

[2] C. Akyel, S. Babic, S. Kincic, and P. Lagacé, "Magnetic force calculation between thin circular coils and thin filamentary circular coil in air," *Journal of Electromagnetic Waves and Applications*, vol. **21**, pp. 1273–1283 (2007).

[3] S. I. Babic and C. Akyel, "Magnetic force calculation between thin coaxial circular coils in air," *Magnetics, IEEE Transactions on*, vol. **44**, pp. 445–452 (2008).

[4] C. Christodoulides, "Comparison of the Ampere and Biot-Savart magnetostatic force laws in their line-current-element forms," *American Journal of Physics*, vol. **56**, pp. 357–362 (1988).

[5] J. C. Maxwell, *A treatise on electricity and magnetism* vol. **1**: Clarendon Press, (1881).

[6] R. Ravaud, G. Lemarquand, S. Babic, V. Lemarquand, and C. Akyel, "Cylindrical magnets and coils: Fields, forces, and inductances," *Magnetics, IEEE Transactions on*, vol. **46**, pp. 3585–3590 (2010).

[7] R. Ravaud, G. Lemarquand, V. Lemarquand, S. Babic, and C. Akyel, "Mutual inductance and force exerted between thick coils," *Progress In Electromagnetics Research*, vol. **102**, pp. 367–380 (2010).

[8] A. Shiri and A. Shoulaie, "A new methodology for magnetic force calculations between planar spiral coils," *Progress In Electromagnetics Research*, vol. **95**, pp. 39–57 (2009).

[9] C. Akyel, S. I. Babic, and M. M. Mahmoudi, "Mutual inductance calculation for non-coaxial circular air coils with parallel axes," *Progress In Electromagnetics Research*, vol. **91**, pp. 287–301 (2009).

[10] S. Babic, F. Sirois, C. Akyel, and C. Girardi, "Mutual inductance calculation between circular filaments arbitrarily positioned in space: alternative to grover's formula," *Magnetics, IEEE Transactions on*, vol. **46**, pp. 3591–3600 (2010).

[11] S. Babic, F. Sirois, C. Akyel, G. Lemarquand, V. Lemarquand, and R. Ravaud, "New formulas for mutual inductance and axial magnetic force between a thin wall solenoid and a thick circular coil of rectangular cross-section," *Magnetics, IEEE Transactions on*, vol. **47**, pp. 2034–2044 (2011).

[12] S. I. Babic and C. Akyel, "Calculating mutual inductance between circular coils with inclined axes in air," *Magnetics, IEEE Transactions on*, vol. **44**, pp. 1743–1750 (2008).

[13] S. I. Babic, F. Sirois, and C. Akyel, "Validity check of mutual inductance formulas for circular filaments with lateral and angular misalignments," *Progress In Electromagnetics Research M*, vol. **8**, pp. 15–26 (2009).

[14] J. T. Conway, "Noncoaxial inductance calculations without the vector potential for axisymmetric coils and planar coils," *Magnetics, IEEE Transactions on*, vol. **44**, pp. 453–462 (2008).

[15] J. T. Conway, "Inductance calculations for circular coils of rectangular cross section and parallel axes using Bessel and Struve functions," *Magnetics, IEEE Transactions on*, vol. **46**, pp. 75–81 (2010).

[16] K. B. Kim, E. Levi, Z. Zabar, and L. Birenbaum, "Mutual inductance of noncoaxial circular coils with constant current density," *Magnetics, IEEE Transactions on*, vol. **33**, pp. 4303–4309 (1997).

[17] A. Benhama, A. Williamson, and A. Reece, "Force and torque computation from 2-D and 3-D finite element field solutions," in *Electric Power Applications, IEE Proceedings*, pp. 25–31 (1999).

[18] J. Coulomb and G. Meunier, "Finite element implementation of virtual work principle for magnetic or electric force and torque computation," *Magnetics, IEEE Transactions on*, vol. **20**, pp. 1894–1896 (1984).

[19] S. Babic and C. Akyel, "Magnetic Force Between Inclined Circular Loops (Lorentz Approach)," *Progress In Electromagnetics Research B*, vol. **38**, pp. 333–349 (2012).

[20] S. Babic and C. Akyel, "Magnetic force between inclined filaments placed in any desired position," *IEEE Trans. Magn.*, vol. **48**, pp. 69–80 (2012).

[21] F. W. Grover, "Inductance calculations," vol. chs. 2 and 13 (1964).

[22] M. Abramowitz and I. A. Stegun, *Handbook of mathematical functions: with formulas, graphs, and mathematical tables* vol. **55**: Dover publications, (1965).

[23] I. S. Gradshtein, I. M. Ryzhik, A. Jeffrey, and D. Zwillinger, *Table of integrals, series, and products:* Academic press, (2007).

[24] S. L. Campbell, J.-P. Chancelier, and R. Nikoukhah, *Modeling and Simulation in SCILAB*: Springer (2010).

Biographies

Anele Amos Onyedikachi received his B.Eng (Hons) degree from the University of Ilorin (UNILORIN) in 2007. He obtained a double-master degree "MTech" and "MSc" in 2012 from the Tshwane University of Technology

(TUT), South Africa and the Ecole Superieure d'Ingenieur en Electronique et Electrotechnique (ESIEE), France respectively. He is currently a PhD student with TUT and the Universite de Versailles St-Quentin-en-Yvelines (UVSQ) in France. He has authored 7 peer-reviewed research papers at international conferences. He is a student member of the IEEE. His research interest is in the field of sustainable engineering, energy and environment.

Yskandar Hamam graduated as a Bachelor of the American University of Beirut (AUB) in 1966. He obtained his M.Sc. in 1970 and Ph.D. in 1972 from the University of Manchester Institute of Science and Technology. He also obtained his "Diplôme d'Habilitation à Diriger des Recherches" (equivalent to D.Sc.) from the « Université des Sciences et Technologies de Lille » in 1998. He conducted research activities and lectured in England, Brazil, Lebanon, Belgium and France. He was the head of the Control department and dean of faculty at ESIEE, France. He was an active member in modelling and simulation societies and was the president of EUROSIM. He was the Scientific Director of the French South African Institute of Technology (F'SATI) at TUT in South Africa from 2007 to 2012. He is currently professor at the Department of Electrical Engineering of TUT. He has authored/co-authored about 300 papers in archival journals and conference proceedings as well as book contributions. He is senior member of the IEEE.

Yasser Alayli received his PhD in applied physics from Pierre and Marie Curie University of Paris, France in 1978. He is professor in the field of engineering sciences and optronics at Versailles University, France. He was director of LISV, UVSQ, Parisfrom 2008 to March, 2013. He is currently the coordinator of a European project "Mobility Motivator" and the head of Move'oTreve "Charging of Electric Vehicles by Magnetic Induction" project. His research interests include precision engineering domain with sub-nanometric accuracy, optical sensors and nanotechnologies.

Karim Djouani is professor, scientist and technical group supervisor of soft computing, telecommunication, networking systems and Robotics. ?Since January 2011 he is Full professor at University Paris Est-Creteil (UPEC), France and Tshwane University of Technology, Pretoria, South Africa. ?From July 2008 to December 2010, he was seconded by the French Ministry of Higher Education to the French South African Institute of Technology (F'SATI) at Tshwane University of Technology (TUT), Pretoria, South Africa. Till July he is also with the SCTIC team of the LISSI lab, University Paris Est. He was also national and European projects manager at the LISSI Lab. His current works focus on the development of novel and highly e?cient algorithms for reasoning systems with uncertainty as well as optimization, for distributed systems, networked control systems, wireless ad-hoc network, wireless and mobile communication, and wireless sensors networks as well as Robotics. He has authored/co-authored over 150 articles in archival journals and conference proceedings as well as ?ve chapters in edited books. ?Prof. Djouani is a Member of IEEE communication and computer societies, Exystenze (European Center of Excellence in Complexity) and several National Research task Group (GDR-MACS, GDR-ISIS)

Investigating the Impacts of Lateral and Angular Misalignments between Circular Filaments

Anele O. Amos[1,2], Hamam Yskandar[1,2,3], Alayli Yasser[2]
and Djouani Karim[1,4]

[1]Dept. of Electrical Engineering, Tshwane University of Technology, Pretoria, South Africa
[2]LISV Laboratory, UVSQ, Paris, France
[3]ESIEE Paris Est University, Paris, France
[4]LISSI Laboratory Paris Est University, Paris, France
anelea@tut.ac.za, hamama@tut.ac.za, yasser.alayli@lisv.uvsq.fr, djouanik@tut.ac.za

Received 24 August 2013; Accepted 18 December 2013
Publication 23 January 2014

Abstract

This paper analyses the impacts of lateral and angular misalignments on the mutual inductance and the magnetic force between circular filaments which are arbitrarily positioned in space. Advanced and relevant models available in the literature are used to accomplish the aim of this paper. Using SCILAB application software, the results obtained based on the theoretical model show that as the coil misalignments increase the values of the mutual inductance and the magnetic force keep decreasing and increasing with respect to certain variable rotation angle at any point of the secondary coil. In order to further investigate the impact of coil separation distance and misalignments on the amount of voltage induced in the secondary coil, a model of air-cored transformer is constructed. The experimental results obtained show that the amount of the induced magnetic flux from the primary coil into the secondary coil becomes weaker if the coil separation distance and coil misalignments increase. As a result, a much smaller value of the mutual inductance is obtained resulting to a much smaller induced electromotive force in the secondary coil.

Journal of Machine to Machine Communications, Vol. 1, 35–56.
doi: 10.13052/jmmc2246-137X.113

In conclusion, this study shows that the full benefits of contactless inductive power transfer (CIPT) systems will not be realized if issues regarding coil separation distance and misalignments are not tackled in the model to be designed for the CIPT transformer.

Keywords: Circular filaments, coil misalignments, mutual inductance, magnetic force.

1 Introduction

In order to reduce the amount of airborne pollution caused by the transportation system, many of the big automobile companies have been compelled to move from the manufacturing of internal combustion engine vehicles (ICEVs) to hybrid electric vehicles (HEVs), hydrogen fuel cell vehicles (HFCVs) and electric vehicles (EVs) [1]. Considering high oil prices and environmental awareness, the development of EVs is considered as a healthier mode of transportation because the electricity they consume could be generated from a wide range of sources which include fossil fuel, nuclear power and renewable sources such as tidal power, solar power, wind power or any combination of those [1, 2]. Although EVs are considered as a favourable solution for a greener energy but, users and owners of EVs feel uncomfortable because EVs require sufficient battery storage onboard to provide sufficient driving autonomy.

To overcome the issue of limited driving distance per charge, conventional wired system e.g. plug-in connectors have been commonly proposed for EV battery charging [1]. Although plug-in connector is a simple and reliable solution however, it may result to safety risks (e.g. electrocution) in wet and damp conditions (see Figures 1 and 2), it is a source of inconvenience (see Fig. 2) and lastly, it only enables stationary charging which means that an EV has to be stationary during the duration of its charge replacement.

Currently, contactless inductive power transfer (CIPT) system is used to overcome the problems of plug-in connectors. CIPT systems are a novel technology used for transferring electrical energy over a relatively large air-gap via high frequency magnetic fields [3]. The potential advantages of CIPT systems over plug-in connectors include an increased driving range, immunity to dirt, dust, water, ice and chemicals, reduced cabling and risk of cable breakage, low maintenance requirement and the use of a high frequency (10 kHz – 150 kHz) magnetic fields which cannot cause electrocution. Wireless transfer of electrical power to the on-board battery storage system of EVs

Figure 1   Plug-in Connector for EV Battery Charging

Figure 2   Plug-in Connector: Exposed Electrical Terminal

is accomplished through the major component of CIPT system which is the CIPT transformer (see Figures 3 and 4).

Although CIPT transformer which consists of air-cored coils plays a major role in CIPT systems however, coil misalignments (e.g. lateral and angular) are their inherent problem. As a result, its full potential is limited because the value of the mutual inductance as well as the magnetic force exerted on the current carrying conductor depends on the relative position and orientations of the coils in the CIPT transformer [4].

Air-cored coils are widely used in various electromagnetic applications. They are preferred to iron-cored type because of their design objectives of controllability and capability of a high power transfer [5]. The authors of

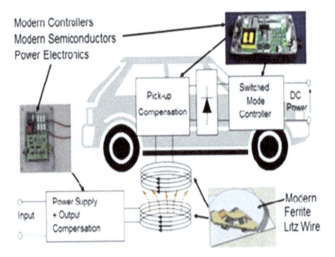

Figure 3   CIPT System for EV Battery Charging

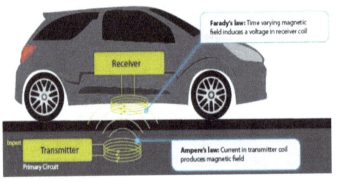

Figure 4   Fundamental Principles of CIPT Systems

this paper consider the use of circular filaments amongst other air-cored coils since in several electromagnetic applications regarding coil misalignments, the optimal magnetic coupling between circular filamentary coils is required [6, 7].

Computations of the mutual inductance and the magnetic force between coaxial circular coils have been completely solved by the authors in [7–20]. However, the present-day focus has been shifted to the computation of the mutual inductance and the magnetic force between circular coils with lateral and angular misalignments [21–30]. Furthermore, these computations can be achieved correctly and speedily by using finite element and boundary element methods [31, 32]. Notwithstanding, the authors in [22, 24] argue that analytical and semi-analytical methods can be used to achieve this task since they significantly simplify the mathematical procedures which in turn leads to a considerable reduction of the computational effort. It is also concluded that the mathematical models formulated for the mutual inductance and the magnetic force between filamentary circular coils are slightly more general and simpler to use (i.e., easy to understand, numerically suitable and easily applicable for engineers and physicists).

Based on this information, the authors of this paper analyse the impacts of lateral and angular misalignments on the mutual inductance and the magnetic force between circular filaments arbitrarily positioned in space based on the advanced and relevant models available in [24]. This task is achieved as follows. Section 2 presents the advanced and relevant models which are obtainable in the literature. In section 3, the theoretical results obtained using SCILAB application software is given as well as its discussion. Section 4 presents the model constructed and its experimental results while section 5 concludes the paper.

2 Advanced and Relevant Mathematical Model

This section presents the relevant and advanced models for computing the mutual inductance and the magnetic force between circular filaments which are arbitrarily positioned in space

2.1 Mutual Inductance Model between Circular Filaments using Magnetic Vector Potential Approach

The mutual inductance between the filamentary circular coils as shown in Fig. 5 can be computed by [24] as

$$M = \frac{\mu_0 R_S}{\pi} \int_0^{2\pi} \frac{[p_1 \cos \phi + p_2 \sin \phi + p_3] \Psi(k)}{k \sqrt{V_0^3}} d\phi \qquad (1)$$

where

$$p_1 = \pm \frac{\gamma c}{l}, \quad p_2 = \mp \frac{\beta l^2 + \gamma ab}{lL} \quad \text{and} \quad p_3 = \frac{\alpha c}{L}$$

$$p_4 = \mp \frac{\beta ab + \gamma l^2 + \delta bc}{lL} \quad \text{and} \quad p_5 = \mp \frac{\beta c - \delta a}{lL}$$

$$\alpha = \frac{R_S}{R_P}, \quad \beta = \frac{x_C}{R_P}, \quad \gamma = \frac{Y_C}{R_P} \quad \text{and} \quad \delta = \frac{Z_C}{R_P}$$

$$L = \sqrt{a^2 + b^2 + c^2} \quad \text{and} \quad l = \sqrt{a^2 + c^2}$$

$$\Psi(k) = \left(1 - \frac{k^2}{2}\right) K(k) - E(k)$$

$$k = \sqrt{\frac{4V_0}{A_0 + 2V_0}}$$

$$V_0{}^2 = \alpha^2 \left[\left(1 - \frac{b^2 c^2}{l^2 L^2}\right) \cos^2 \phi + \frac{c^2}{l^2} \sin^2 \phi + \frac{abc}{l^2 L} \sin \phi\right] +$$

$$\beta^2 + \gamma^2 \mp 2\alpha \frac{\beta ab - \gamma l^2}{lL} \cos \phi \mp \frac{2\alpha \beta c}{l} \sin \phi$$

$$A_0 = 1 + \alpha^2 + \beta^2 + \gamma^2 + \delta^2 + 2\alpha(p_4 \cos \phi + p_5 \sin \phi)$$

where μ_0 is the magnetic permeability of space, R_P is the radius of primary coil, R_S is the radius of the secondary coil, α is the shape factor of the circular coil, $(x_C, y_C \text{ and } z_C)$ is centre of the secondary coil, a, b and c are the parameters defining the secondary coil plane λ, k is a variable and not indices, $K(k)$ and $E(k)$ are the complete integral of the first and second kind respectively [33, 34].

2.2 Magnetic Force between Circular Filaments using Mutual Inductance Approach

The magnetic force f_g between filamentary circular coils arbitrarily positioned in space can be computed by [14]

$$F_g = I_P I_S \frac{\partial M}{\partial g} \tag{2}$$

where I_P and I_S are the primary and secondary currents in the coil, M is the mutual inductance given in (1) and g = x, y, or z are the xyz components.

Finding the first derivative in (2), the magnetic force can be obtained by the following components:

$$F_X = \frac{\mu_0 \alpha I_P I_S}{\pi} \int_0^{2\pi} I_X d\phi$$

$$F_Y = \frac{\mu_0 \alpha I_P I_S}{\pi} \int_0^{2\pi} I_Y d\phi \qquad (3)$$

$$F_Z = \frac{\mu_0 \alpha I_P I_S}{4\pi} \int_0^{2\pi} I_Z d\phi$$

where

$$I_x = q_3 \sin\phi \frac{\Psi(k)}{k\sqrt{V_0^3}} - \frac{kT_0(p_1 \cos\phi + p_2 \sin\phi + p_3)}{8\sqrt{V_0^9}} \times$$

$$\left\{ \Psi(k) \left[[A_0 - 2V_0{}^2] + \frac{12V_0}{k_2} \right] - \Phi(k) \frac{k^2 [A_0 - 2V_0{}^2]}{2} \right\}$$

$$I_y = (-q_5 \cos\phi + q_4 \sin\phi) \frac{\Psi(k)}{k\sqrt{V_0^3}} - \frac{kS_0(p_1 \cos\phi + p_2 \sin\phi + p_3)}{8\sqrt{V_0^9}} \times$$

$$\left\{ \Psi(k) \left[[A_0 - 2V_0{}^2] + \frac{12V_0}{k^2} \right] - \Phi(k) \frac{k^2 [A_0 - 2V_0{}^2]}{2} \right\}$$

$$I_Z = \frac{[p_1 \cos\phi + p_2 \sin\phi + p_3] L_0}{\sqrt{V_0^5}} \Theta(k)$$

where

$$q_1 = \frac{\gamma l^2 - \beta ab}{lb}, \ q_2 = -\frac{\beta c}{l}, \ q_3 = -\frac{l}{L}, \ q_4 = \frac{ab}{lL}, \ q_5 = -\frac{c}{l}$$

$$L = \sqrt{a^2 + b^2 + c^2} \ and \ l = \sqrt{a^2 + c^2}$$

$$\alpha = \frac{R_S}{R_P}, \beta = \frac{x_C}{R_P}, \gamma = \frac{y_C}{R_P}, \delta = \frac{z_C}{R_P}$$

$$p_1 = \pm\frac{\gamma c}{l}, \ p_2 = \mp\frac{\beta l^2 + \gamma ab}{lL}, \ p_3 = \frac{\alpha c}{L}$$

$$p_4 = \mp \frac{\beta ab - \gamma l^2 \delta bc}{lL}, \quad p_5 = \mp \frac{\beta c - \delta a}{lL}$$

$$\Psi(k) = \left(1 - \frac{k^2}{2}\right) K(k) - E(k)$$

$$\Phi(k) = \frac{E(k)}{1 - k^2} - K(k)$$

$$\Theta(k) = k \left\{ \Psi(k) - \frac{k^2}{2} \Phi(k) \right\}$$

$$k = \sqrt{\frac{4V_0}{A_0 + 2V_0}},$$

$$V_0^2 = \beta^2 + \gamma^2 + \alpha^2 (l_1 \cos^2\phi + l_2 \sin^2\phi + l_3 \sin 2\phi) +$$
$$2\alpha(q_1 \cos\phi + q_2 \sin\phi)$$

$$A_0 = 1 + \alpha^2 + \beta^2 + \gamma^2 + \delta^2 + 2\alpha(p_4 \cos\phi + p_5 \sin\phi)$$

$$T_0 = \beta + \alpha \left[q_4 \cos\phi + q_5 \sin\phi \right]$$

$$S_0 = \gamma - \alpha q_3 \cos\phi$$

$$L_0 = \delta + \alpha \left[q_6 \cos\phi - q_7 \sin\phi \right]$$

$$q_6 = -\frac{bc}{lL}, \quad q_7 = -\frac{a}{l}$$

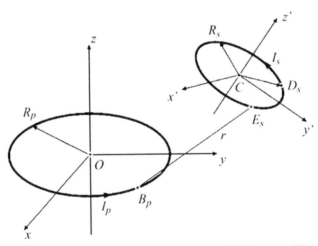

Figure 5   Circular Filaments with arbitrary misalignments [24]

3 Theoretical Result Obtained

In order to investigate the impacts of coil misalignments between circular coils arbitrarily positioned in space, the authors of this paper re-stated equations (1) and (3) in terms of the geometric configurations and common notations for circular filaments with arbitrary lateral and angular as given in [24]. Shown in Figures 6–13 are the results obtained using SCILAB application software [35] and the data used is shown in Table 1.

Table 1   Data used for Simulation [24]

Variable Parameters			
x_1 (m)	x_2 (m)	P (m)	θ (degree)
0.05	0.0125	0.04	30
0.10	0.0250	0.08	45
0.15	0.0375	0.14	60
0.20	0.0500	0.19	75
Constant Parameters			
R_P	0.16m		
R_S	0.10m		
μ_0	$4\pi \times 10^{-7} H/m$		

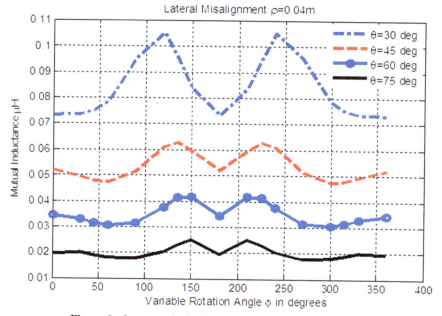

Figure 6   Impacts of Misalignments on the Mutual Inductance

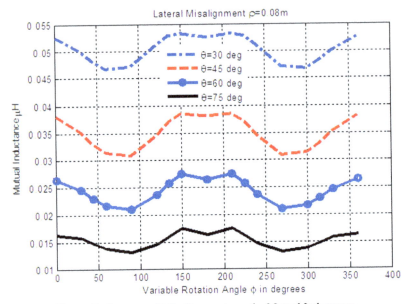

Figure 7   Impacts of Misalignments on the Mutual Inductance

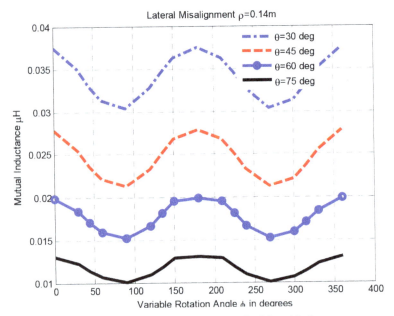

Figure 8   Impacts of Misalignments on the Mutual Inductance

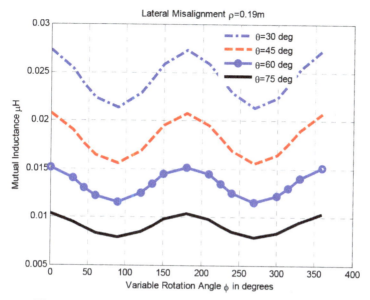

Figure 9   Impacts of Misalignments on the Mutual Inductance

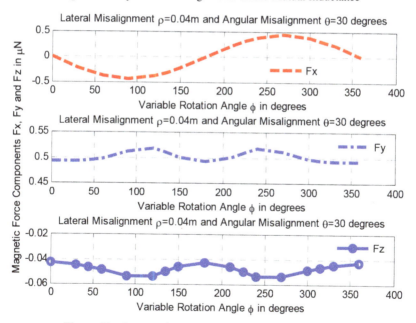

Figure 10   Impact of Misalignments on the Magnetic Force

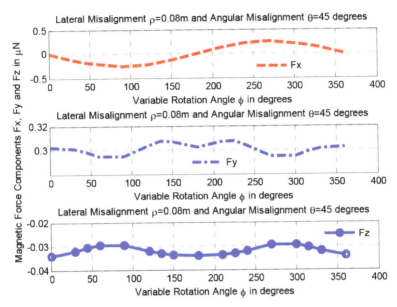

Figure 11   Impact of Misalignments on the Magnetic Force

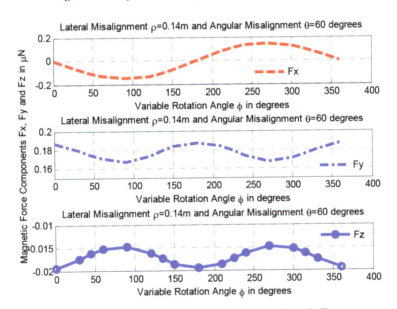

Figure 12   Impacts of Misalignments on the Magnetic Force

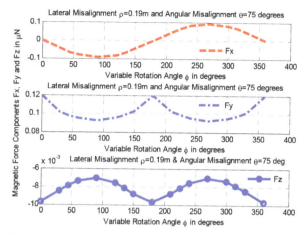

Figure 13   Impacts of Misalignments on the Magnetic Force

3.1 Discussion of the Theoretical Results

The results obtained in Figures 6–13 show the impacts of lateral and angular misalignments on the mutual inductance and the magnetic force between circular filaments arbitrarily positioned in space. These results are obtained by plotting the mutual inductance and the magnetic force against the variable rotation angle at any point of the secondary coil. Based on the data given in Table 1, it is clearly seen that the values of the mutual inductance (see Figures 6–9) and the magnetic force components (see Figures 10–13) keep decreasing and increasing at certain variable rotation angle.

4 Air-Cored Transformer and Experimental Results

The primary and secondary coils of transformers possess mutual inductance when they are magnetically linked together by a common magnetic flux. Hence, mutual inductance is an important operating property of transformers. Notwithstanding, its amount depends very much on the coil separation distance as well as lateral and angular misalignments. This implies that the amount of induced magnetic flux from the primary coil into the secondary coil is weaker if the coil separation distance and coil misalignments increase. As a result, a much smaller value of the mutual inductance is obtained resulting to a much smaller induced electromotive force (emf) in the secondary coil.

Based on this information, in order to further investigate the impact of coil separation distance as well as coil misalignments, a model of air-cored

Figure 14   AC/DC Variable Power Supply, Multi-meter and Coreless Transformer

Figure 15   Author Investigating Impacts of Coil Separation Distance and Misalignments

transformer which consists of rectangular coils is constructed. Shown in Figures 14–16 is the prototype set up which consists of variable AC/DC power supply, multi-meter and the coreless transformer.

The primary and the secondary air-cored rectangular coils are of the same dimension and number of turns. In order to determine the amount of electromotive force (which is expressed in volts) induced in the secondary

Figure 16   Colleagues during Prototype Setup

coil, ac voltage of one volt is supplied to the primary coil of the coreless transformer at different coil separation distance and misalignments.

4.1 Experimental Measurements

Shown in Tables 2–4 are the experimental results obtained for the air-cored transformer which consists of rectangular coils. The experimental measurements show that as the coil separation distance as well as the lateral and angular misalignments increase the amount of the induced magnetic flux from the primary coil into the secondary coil becomes weaker. As a result, a much smaller value of the mutual inductance is obtained resulting to a much smaller induced emf (which is expressed in volts) in the secondary coil.

Table 2 Data used for Simulation		
Constant Parameters		
Primary and Secondary Inductance $L_P = L_S$	$3.2\ mH$	Primary Current I_P=0.995 $A \approx 1\ A$
$X_L = 2\pi F L$	1.00544Ω	Coil Separation Distance = $10\ mm,\ 20\ mm,\ 30\ mm,\ 40\ mm\ \&\ 50\ mm$
Primary Voltage V_P	$1V$	Number of Turns $L_P = L_S$= 50

Table 3   Coil Separation Distance of 10 mm

Angular Misalignment θ (degree)	Lateral Misalignment d (mm)	Induced emf in the Secondary Coil V_S (Volts)
0	0.00	0.550
10	0.05	0.505
30	0.10	0.456
45	0.15	0.401

Table 4   Coil Separation Distance of 20 mm

Angular Misalignment θ (degree)	Lateral Misalignment d (mm)	Induced emf in the Secondary Coil V_S (Volts)
0	0.00	0.385
10	0.05	0.353
30	0.10	0.320
45	0.15	0.292

Table 5   Coil Separation Distance of 30 mm

Angular Misalignment θ (degree)	Lateral Misalignment d (mm)	Induced emf in the Secondary Coil V_S (Volts)
0	0.00	0.278
10	0.05	0.246
30	0.10	0.211
45	0.15	0.204

Table 6   Coil Separation Distance of 40 mm

Angular Misalignment θ (degree)	Lateral Misalignment d (mm)	Induced emf in the Secondary Coil V_S (Volts)
0	0.00	0.195
10	0.05	0.170
30	0.10	0.136
45	0.15	0.113

Table 7   Coil Separation Distance of 50 mm

Angular Misalignment θ (degree)	Lateral Misalignment d (mm)	Induced emf in the Secondary Coil V_S (Volts)
0	0.00	0.098
10	0.05	0.083
30	0.10	0.082
45	0.15	0.075

5 Conclusion

The authors of this paper analyse the impacts of lateral and angular misalignments on the mutual inductance and the magnetic force between circular filaments which are arbitrarily positioned in space. Advanced and relevant models available in the literature are used to achieve the theoretical results obtained in Figures 6–13. The simulations obtained show that as the coil misalignments increase the values of the mutual inductance and the magnetic force keep decreasing and increasing with respect to certain variable rotation angle at any point of the secondary coil.

In order to further investigate the impact of coil separation distance as well as lateral and angular misalignments on the amount of voltage induced in the secondary coil, a model of air-cored transformer which consists of rectangular coils is constructed. The experimental results obtained show that the amount of the induced magnetic flux from the primary coil into the secondary coil becomes weaker if the coil separation distance and coil misalignments increase. As a result, a much smaller value of the mutual inductance is obtained resulting to a much smaller induced electromotive force in the secondary coil.

In conclusion, this study shows that the full benefits of contactless inductive power transfer (CIPT) systems will not be realized if issues regarding coil separation distance and misalignments are not tackled in the model to be designed for the CIPT transformer.

References

[1] H. H. Wu, A. Gilchrist, K. Sealy, P. Israelsen, and J. Muhs, "A review on inductive charging for electric vehicles," in *Electric Machines & Drives Conference (IEMDC), (2011) IEEE International*, pp. 143–147 (2011).

[2] J. Meins, G. Bühler, R. Czainski, and F. Turki, "Contactless inductive power supply," *Download this and any more recent reports using Google*, (2006).

[3] C.-S. Wang, O. H. Stielau, and G. A. Covic, "Design considerations for a contactless electric vehicle battery charger," *Industrial Electronics, IEEE Transactions on*, vol. **52**, pp. 1308–1314 (2005).

[4] K. Fotopoulou and B. W. Flynn, "Wireless power transfer in loosely coupled links: Coil misalignment model," *Magnetics, IEEE Transactions on*, vol. **47**, pp. 416–430 (2011).

[5] C. S. Wang, O. H. Stielau, and G. A. Covic, "Load models and their application in the design of loosely coupled inductive power transfer

systems," in Power System Technology, 2000. *Proceedings. PowerCon 2000. International Conference on*, pp. 1053–1058 (2000).

[6] J. Acero, C. Carretero, I. Lope, R. Alonso, O. Lucia, and J. M. Burdio, "Analysis of the Mutual Inductance of Planar-Lumped Inductive Power Transfer Systems," *Industrial Electronics, IEEE Transactions on*, vol. **60**, pp. 410–420 (2013).

[7] S. Babic and C. Akyel, "New mutual inductance calculation of the magnetically coupled coils: Thin disk coil-thin wall solenoid," *Journal of Electromagnetic Waves and Applications*, vol. **20**, pp. 1281–1290 (2006).

[8] S. Babic, F. Sirois, C. Akyel, G. Lemarquand, V. Lemarquand, and R. Ravaud, "New formulas for mutual inductance and axial magnetic force between a thin wall solenoid and a thick circular coil of rectangular cross-section," *Magnetics, IEEE Transactions on*, vol. **47**, pp. 2034–2044 (2011).

[9] S. I. Babic and C. Akyel, "Magnetic force calculation between thin coaxial circular coils in air," *Magnetics, IEEE Transactions on*, vol. **44**, pp. 445–452 (2008).

[10] S. Butterworth, "LIII. On the coefficients of mutual induction of eccentric coils," *The London, Edinburgh, and Dublin Philosophical Magazine and Journal of Science*, vol. **31**, pp. 443–454 (1916).

[11] C. Christodoulides, "Comparison of the Ampere and Biot-Savart magnetostatic force laws in their line-current-element forms," *American Journal of Physics*, vol. **56**, pp. 357–362 (1988).

[12] H. B. Dwight, *Electrical coils and conductors*: McGraw-Hill, (1945).

[13] F. W. Grover, "The calculation of the mutual inductance of circular filaments in any desired positions," *Proceedings of the IRE*, vol. **32**, pp. 620–629 (1944).

[14] F. W. Grover, "Inductance calculations," vol. chs. 2 and 13 (1964).

[15] P. Kalantarov and L. Teitlin, "Inductance Calculations," *Bucharest: Tehnica.–1958* (1955).

[16] J. C. Maxwell, *A treatise on electricity and magnetism* vol. **1**: Clarendon Press, (1881).

[17] R. Ravaud, G. Lemarquand, S. Babic, V. Lemarquand, and C. Akyel, "Cylindrical magnets and coils: Fields, forces, and inductances," *Magnetics, IEEE Transactions on*, vol. **46**, pp. 3585–3590 (2010).

[18] R. Ravaud, G. Lemarquand, V. Lemarquand, S. Babic, and C. Akyel, "Mutual inductance and force exerted between thick coils," *Progress In Electromagnetics Research*, vol. **102**, pp. 367–380 (2010).

[19] A. Shiri and A. Shoulaie, "A new methodology for magnetic force calculations between planar spiral coils," *Progress In Electromagnetics Research*, vol. **95**, pp. 39–57 (2009).

[20] C. Snow, *Formulas for computing capacitance and inductance* vol. **544**: US Govt. Print. Off., (1954).

[21] C. Akyel, S. I. Babic, and M. M. Mahmoudi, "Mutual inductance calculation for non-coaxial circular air coils with parallel axes," *Progress In Electromagnetics Research*, vol. **91**, pp. 287–301 (2009).

[22] S. Babic and C. Akyel, "Magnetic force between inclined filaments placed in any desired position," *IEEE Trans. Magn.*, vol. **48**, pp. 69–80 (2012).

[23] S. Babic and C. Akyel, "Magnetic Force Between Inclined Circular Loops (Lorentz Approach)," *Progress In Electromagnetics Research B*, vol. **38**, pp. 333–349 (2012).

[24] S. Babic, F. Sirois, C. Akyel, and C. Girardi, "Mutual inductance calculation between circular filaments arbitrarily positioned in space: alternative to grover's formula," *Magnetics, IEEE Transactions on*, vol. **46**, pp. 3591–3600 (2010).

[25] S. I. Babic, C. Akyel, and M. M. Mahmoudi, "Mutual Inductance Calculation between Circular Coils with Lateral and Angular Misalignment," *Session 3AP*, p. 156 (2009).

[26] S. I. Babic, C. Akyel, Y. Ren, and W. Chen,"Magnetic force calculation between circular coils of rectangular cross section with parallel axes for superconducting magnet," *Progress In Electromagnetics Research B*, vol. **37**, pp. 275–288 (2012).

[27] J. Conway, "Forces between thin coils with parallel axes using bessel functions," *Private Communication*, Jun, (2011).

[28] J. T. Conway, "Noncoaxial inductance calculations without the vector potential for axisymmetric coils and planar coils," *Magnetics, IEEE Transactions on*, vol. **44**, pp. 453–462 (2008).

[29] J. T. Conway, "Inductance calculations for circular coils of rectangular cross section and parallel axes using Bessel and Struve functions," *Magnetics, IEEE Transactions on*, vol. **46**, pp. 75–81 (2010).

[30] Y. Ren, "Magnetic force calculation between misaligned coils for a superconducting magnet," *Applied Superconductivity, IEEE Transactions on*, vol. **20**, pp. 2350–2353 (2010).

[31] A. Benhama, A. Williamson, and A. Reece, "Force and torque computation from 2-D and 3-D finite element field solutions," in *Electric Power Applications, IEE Proceedings-*, pp. 25–31 (1999).

[32] J. Coulomb and G. Meunier, "Finite element implementation of virtual work principle for magnetic or electric force and torque computation," *Magnetics, IEEE Transactions on*, vol. **20**, pp. 1894–1896 (1984).

[33] M. Abramowitz and I. A. Stegun, *Handbook of mathematical functions: with formulas, graphs, and mathematical tables* vol. **55**: Dover publications, (1965).

[34] I. S. Gradshtein, I. M. Ryzhik, A. Jeffrey, and D. Zwillinger, *Table of integrals, series, and products*: Academic press, (2007).

[35] S. L. Campbell, J.-P. Chancelier, and R. Nikoukhah, *Modeling and Simulation in SCILAB*: Springer (2010).

Biographies

Anele Amos Onyedikachi received his B.Eng (Hons) degree from the University of Ilorin (UNILORIN) in 2007. He obtained a double-master degree "MTech" and "MSc" in 2012 from the Tshwane University of Technology (TUT), South Africa and the Ecole Superieure d'Ingenieur en Electronique et Electrotechnique (ESIEE), France respectively. He is currently a PhD student with TUT and the Universite de Versailles St-Quentin-en-Yvelines (UVSQ) in France. He has authored 7 peer-reviewed research papers at international conferences. He is a student member of the IEEE. His research interest is in the field of sustainable engineering, energy and environment.

Yskandar Hamam graduated as a Bachelor of the American University of Beirut (AUB) in 1966. He obtained his M.Sc. in 1970 and Ph.D. in 1972 from the University of Manchester Institute of Science and Technology. He also obtained his "Diplôme d'Habilitation à Diriger des Recherches" (equivalent to D.Sc.) from the « Université des Sciences et Technologies de Lille » in 1998. He conducted research activities and lectured in England, Brazil, Lebanon, Belgium and France. He was the head of the Control department and dean of faculty at ESIEE, France. He was an active member in modelling and simulation societies and was the president of EUROSIM. He was the Scientific Director of the French South African Institute of Technology (F'SATI) at TUT in South Africa from 2007 to 2012. He is currently professor at the Department of Electrical Engineering of TUT. He has authored/co-authored about 300 papers in archival journals and conference proceedings as well as book contributions. He is senior member of the IEEE.

Yasser Alayli received his PhD in applied physics from Pierre and Marie Curie University of Paris, France in 1978. He is professor in the field of engineering sciences and optronics at Versailles University, France. He was director of LISV, UVSQ, Paris from 2008 to March, 2013. He is currently the coordinator of a European project "Mobility Motivator" and the head of Move'oTreve

"Charging of Electric Vehicles by Magnetic Induction" project. His research interests include precision engineering domain with sub-nanometric accuracy, optical sensors and nanotechnologies.

Karim Djouani is professor, scientist and technical group supervisor of soft computing, telecommunication, networking systems and Robotics. Since January 2011 he is Full professor at University Paris Est-Creteil (UPEC), France and Tshwane University of Technology, Pretoria, South Africa. ?From July 2008 to December 2010, he was seconded by the French Ministry of Higher Education to the French South African Institute of Technology (F'SATI) at Tshwane University of Technology (TUT), Pretoria, South Africa. Till July he is also with the SCTIC team of the LISSI lab, University Paris Est. He was also national and European projects manager at the LISSI Lab. ?His current works focus on the development of novel and highly e?cient algorithms for reasoning systems with uncertainty as well as optimization, for distributed systems, networked control systems, wireless ad-hoc network, wireless and mobile communication, and wireless sensors networks as well as Robotics. He has authored/co-authored over 150 articles in archival journals and conference proceedings as well as ?ve chapters in edited books. ?Prof. Djouani is a Member of IEEE communication and computer societies, Exystenze (European Center of Excellence in Complexity) and several National Research task Group (GDR-MACS, GDR-ISIS).

Bayesian Updating of the Gamma Distribution for the Analysis of Stay Times in a System

K. Aboura and Johnson I. Agbinya

*College of Business Administration, University of Dammam, Dammam, Saudi
Arabia, kaboura@ud.edu.sa*
Electronic Engineering, Latrobe University, Melbourne, Australia
J.Agbinya@latrobe.edu.au

Received 4 July 2013; Accepted 15 November 2013
Publication 23 January 2014

Abstract

The evaluation of traffic in a system is an important measurement in many
studies. Counting the number of items in a system has applications in all
processing operations. Electronic messages circulating in a network, clients
shopping in a supermarket and students attending programs in a school are
examples of entities entering, staying and exiting a system. We introduce a
Bayesian updating methodology for the gamma distribution for the analysis
of stay times in a system. The methodology was first developed for areas
monitored by surveillance cameras. The number of people in the covered
area was determined and the average stay time was estimated using a gamma
probability distribution. We extend the application to the generic case and
present a simple updating methodology for the estimation of the model
parameters.

Keywords: Traffic estimation, gamma distribution, Bayesian statistics.

1 Introduction

Counting the number of entities in a system is essential for a multitude of
management and monitoring functions. For example, forecasting the number
of students in a school is essential to the planning of all functions in the

Journal of Machine to Machine Communications, Vol. 1 , 57–72.
doi: 10.13052/jmmc2246-137X.114

school. Students register, follow programs and at times drop from the school before completing their degrees. When the school size is large, it becomes essential to estimate the flow of students and regress it on capacity, staffing and funding variables to forecast needed resources. In general, it is often the case that management desires to estimate stay times in a system. We introduce a Bayesian updating methodology for the gamma distribution of stay times in a system. The methodology was initially developed for counting people in areas monitored by surveillance cameras [1]. The method was useful in the development of technology to estimate traffic through real time video information processing. The number of people in the covered area was determined and the average stay time was estimated using a gamma probability distribution model. A Bayesian updating methodology was used. We extend the application to the generic case and present a simple updating methodology for the estimation of the probability model parameters.

2 The Number of Entities in the System

Although there is often an error associated with the assessment of the number of entities in a system, one can achieve a high reliability in many situations. In the case of a school, the exact number of students can be obtained through registration records. If the entities are students in a school or people in some geographic area, the count results in $N(t)$, the number of entities in the system at time t (Figure 1). The units of time in Figure 1 are study dependent. They can

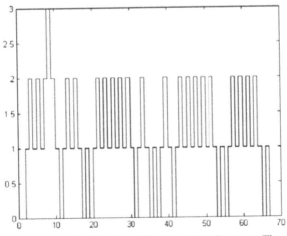

Figure 1 Nt - the Number of Entities in the System at Time t

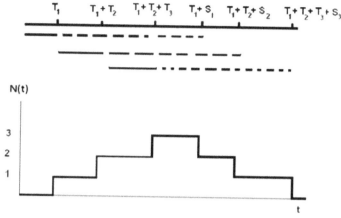

Figure 2 $N(t)$ is the compounding of all processes

be frames per second, months or semesters. $N(t)$ is a stochastic process, the compounding of all processes representing the arrival, stay and departure of entities in the system. For simplicity, we will use the example of people being counted in a geographic area. Starting a cycle at time $t = 0$, where $N(t) = 0$, at time T_1 the first person arrives to the system and stays a time S_1. The second person arrives at time $T_1 + T_2$, $T_2 > 0$ and stays a time S_2 etc. $T_1, T_2, T_3 \ldots$ are the inter-arrival times. $S_1, S_2, S_3 \ldots$ are the times spent in the system (Figure 2).

In [1], depending on the location of system, we found that the probabilistic nature of the Ti's and Si's differed according to the time of day, day of the week and season. This observation required the analysis of all data and their separation into classes of homogeneity. This data stratification is necessary in most studies and leads to results for the system in different states. Systems are often complex and can show chaotic behaviour as opposed to a steady state behaviour. An analysis of the system must take into account different periods of homogeneity in the data.

3 Incoming Flow and the Arrival Process

In [1], we derived probability models for the arrival process and the time spent by each person in the system. We used homogeneous data. The probability models were applied within periods of time where the data showed the same probabilistic behaviour. Using prior knowledge and the results of statistical analyses, we classified the data into periods of homogeneity. Considering the incoming flow of the arrival process, we applied probability models and estimated the parameters. Based on preliminary studies, we observed that the T_i's had exponential distributions with mean θ^{-1} (Figure 3). This, in addition to other assumptions, implied that the arrival stochastic process was a Poisson process. The more general candidate for the arrival process was the non-homogeneous Poisson process (NHPP) [2] where θ varies with time and is not a constant. However, we restricted ourselves to homogeneity periods of time where the inter-arrival times are considered independent and identically distributed (IID). Many standard techniques can be used to estimate θ. Using the inter-arrival times and their distribution $T_i \sim \text{Exp}(\theta)$, one can obtain an accurate estimate of θ that gets refined with time. To conduct such an analysis, we used a Bayesian approach with a conjugate gamma prior distribution for θ. The use of a conjugate prior leads to a posterior gamma distribution. We found the approach to be effective in the estimation of the parameter θ. While a number of techniques can be used to estimate the mean of an exponential distribution, we preferred the Bayesian approach for its probabilistic estimation of the parameter θ. The posterior gamma distribution of θ offers probability intervals surrounding the posterior mode as an estimate of the inverse of the mean inter-arrival. These probability intervals can be used effectively in studies that conduct sensitivity analyses when the system shows important variability. Often such studies involve simulation for the prediction

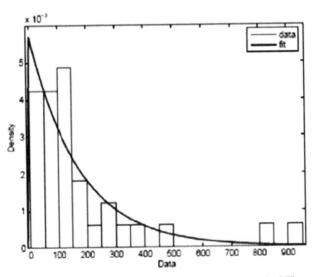

Figure 3   Exponential Distribution Fit for the Inter-arrival Times

of complex situations that cannot be handled analytically. The input to such simulations includes the probability models of inter-arrival times. The use of a point estimate for θ without adequate probability bounds may limit the simulation results. In our approach, we use the assumption of exponential inter-arrival times with mean θ^{-1}, with a prior gamma distribution for θ.

4 Time Spent in the System

$S_1, S_2, S_3 \ldots$ are the times spent in the system. At times, we cannot observe the Si's directly unless we track each person. In other cases, we can use existing records. In the case of video surveillance, the image processing task provides only $N(t)$, the number of people in the system at time t. Using $N(t)$ we determine exact estimates of the average time spent in the system and probability models for the Si's. It can be shown that for a cycle, where $N(t)$ start at $N(t) = 0$ and returns to $N(t) = 0$, that

$$\int_{t_1}^{t_2} N(t)dt = \sum_{i=1}^{n} S_i \tag{1}$$

where n is the number of people who arrived to the system between t_1 and t_2. By collecting the statistic $\sum_{i=1}^{n} S_i$ for all cycles of the same homogeneous time period and taking the average over the total number of people, we obtain the exact sample average time spent in the system by a person and an excellent estimator of the time spent in the system. Often, this estimate is enough for simple purposes. To conduct an inference and predict, one must derive probability models. $\sum_{i=1}^{n} S_i$ is observed for all cycles. Consider the variable $Y = \sum_{i=1}^{n} S_i$. Using the collected data $(Y_1, Y_2, Y_3 \ldots)$ for the same n, and assuming a period of homogeneity, that is the data are IID, we can fit a probability distribution to the Y_i's. From such distribution, one would derive the distribution of the S_i's. For example, if the Si's are gamma distributed then the sum $\sum_{i=1}^{n} S_i$ will be gamma distributed. As it turns out, our study has shown that the data fit a gamma distribution. However, this approach requires collecting considerable amount of data, as one needs large samples $(Y_1, Y_2, \ldots)$ having all the same n number of people that have entered and stayed in the system. Instead, we prefer to use a Bayesian approach that makes use of all collected cycle data. Let $(Y_1, Y_2, \ldots, Y_m)$ be the collected statistics for m cycles, where

$$Y_i = \sum_{j=1}^{n_i} S_{i,j}, i = 1, 2, ..., m \tag{2}$$

Using $(Y_1, Y_2, \ldots, Y_m)$, we want to assess the probability distribution of $S_{i,j}$, the time spent in the system by a person. Based on probability model assumptions, we assume that the $\{S_{i,j}, i, j = 1, 2, \ldots\}$ are independent random variables that are identically distributed. $S_{i,j} \sim$ Gamma(α, λ), $i, j = 1, 2, \ldots$. That is

$$f_{S_{i,j}}(x) = \frac{x^{\alpha-1}\lambda^{\alpha}e^{-\lambda x}}{\Gamma(\alpha)} \tag{3}$$

We want to calculate

$$p(S_{i,j}|y_1, y_2, \ldots, y_m) \tag{4}$$

where $(y_1, y_2, \ldots, y_m)$ are the realizations of $(Y_1, Y_2, \ldots, Y_m)$. Using the Chapman-Kolmogorov equation or the Law of Total Probability, we condition and average over all possible values of α and λ. To do so, we use probability models for these two parameters and make a few model simplifying assumptions. Let $(\alpha_1, \alpha_2, \ldots, \alpha_K)$ be the most likely values for the shape parameter α of the Gamma (α, λ) distribution of $S_{i,j}$. Let Gamma (a, b) be the distribution of the scale parameter λ. It is a natural conjugate prior distribution. Having discretized α, let $p(\alpha)$ be the ensuing discrete prior distribution. If prior information is available on α, it can be used to construct the discrete distribution $p(\alpha)$, either directly or through an Expert Opinion procedure [3], [4]. Otherwise, a flat discrete prior can be used, that is the Uniform distribution over the set $(\alpha_1, \alpha_2, \ldots, \alpha_K)$. Further assume that, to start the procedure, α and λ are independent. This assumption is not a strong one, as the two parameters α and λ do not remain independent long once the data is used. We then have $p(S_{i,j} \mid y_1, y_2, \ldots, y_m)$ equal to

$$\sum_{\alpha} \int_{\lambda} p(S_{i,j}|\alpha, \lambda, y_1, \ldots, y_m)p(\alpha, \lambda|y_1, \ldots, y_m)d\lambda \tag{5}$$

Given (α, λ), $S_{i,j}$ is conditionally independent of $(y_1, y_2, \ldots, y_m)$ and is the Gamma (α, λ) distribution. It remains to evaluate $p(\alpha, \lambda| y_1, y_2, \ldots, y_m)$, the posterior distribution of (α, λ). Using Bayes theorem,

$$p(\alpha, \lambda|y_1, \ldots, y_m) = \frac{1}{\delta}p(y_1, \ldots, y_m|\alpha, \lambda)p(\alpha, \lambda) \tag{6}$$

$$p(\alpha, \lambda|y_1, \ldots, y_m) = \frac{1}{\delta}\prod_{i=1}^{m} p(y_i|\alpha, \lambda)p(\alpha)p(\lambda) \tag{7}$$

where δ is the normalizing factor,

$$\delta = \sum_{\alpha} \int_{\lambda} \prod_{i=1}^{m} p(y_i|\alpha, \lambda)p(\alpha)p(\lambda)d\lambda \tag{8}$$

$$\delta = \sum_\alpha \int_\lambda \prod_{i=1}^m \frac{y_i^{\alpha n_i - 1} \lambda^{\alpha n_i} e^{-\lambda y_i}}{\Gamma(\alpha n_i)} p(\alpha) \frac{\lambda^{a-1} b^a e^{-b\lambda}}{\Gamma(a)} d\lambda \tag{9}$$

$$\delta = \sum_\alpha \int_\lambda \frac{\Gamma(a + \alpha \sum_{i=1}^m n_i)}{\Gamma(a) \prod_{i=1}^m \Gamma(\alpha n_i)} \frac{\prod_{i=1}^m y_i^{\alpha n_i - 1} b^a}{(b + \sum_{i=1}^m y_i)^{a + \alpha \sum_{i=1}^m n_i}} \times \tag{10}$$

$$\frac{\lambda^{(a + \alpha \sum_{i=1}^m n_i) - 1} (b + \sum_{i=1}^m y_i)^{a + \alpha \sum_{i=1}^m n_i} e^{-\lambda(b + \sum_{i=1}^m y_i)}}{\Gamma(a + \alpha \sum_{i=1}^m n_i)} p(\alpha) d\lambda \tag{11}$$

$$\delta = \sum_\alpha \frac{\Gamma(a + \alpha \sum_{i=1}^m n_i)}{\Gamma(a) \prod_{i=1}^m \Gamma(\alpha n_i)} \frac{\prod_{i=1}^m y_i^{\alpha n_i - 1} b^a}{(b + \sum_{i=1}^m y_i)^{a + \alpha \sum_{i=1}^m n_i}} p(\alpha) \tag{12}$$

Since Y_i is the sum of the independent gamma random variables $S_{i,j}$ as defined in Equation 4.2, then $Y_i \sim$ Gamma $(\alpha n_i, \lambda)$. δ is computed in a non-expensive K large summation.

5 Example

The inter-arrival times T_1, T_2, T_3 . . . are obtained from the process $N(t)$. We used several data sets separately for the estimation of θ, the inverse of the mean of the inter-arrival times. The prior distribution on θ is a Gamma (a, b). We varied the prior input and observed the behaviour of the posterior distribution as a function of the choice of the parameters (a, b). The posterior gamma distribution responded robustly to the prior input. In Table 1, using data set 1, we observe the prior mean and corresponding standard deviation followed by the posterior mode and posterior mean and the actual sample mean or empirical mean of θ, for different prior distributions. This is standard Bayesian conjugate analysis of the parameter of an exponential distribution, namely θ, which is well known and well documented. It is an efficient procedure that provides probability bounds rather than just a point estimate. Such probability intervals can be used effectively as input to simulation studies for example where sensitivity analysis is a must. In our case, we provide the methodology for the arrival process in addition to the

Table 1   Bayesian Estimation of The prior distribution on θ Using Data Set 1

a	b	Prior Mean	Prior Standard Deviation	Posterior Mode	Posterior Mean	Empirical Mean
10	900	.0111	.003	.0062	.0064	.0057
5	900	.0055	.0024	.0054	.0057	.0057
3	900	.0033	.0019	.0051	.0054	.0057
2	900	.002	.0016	.0055	.0052	.0057
1	900	.0011	.0011	.0049	.0051	.0057
5	90	.05	.024	.0062	.0064	.0057
3	90	.0333	.019	.0059	.0061	.0057
2	90	.022	.0157	.0057	.0059	.0057
1	90	.0111	.0111	.00557	.0058	.0057

Table 2   Bayesian Estimation of The prior distribution on θ Using Data Set 2

a	b	Prior Mean	Prior Standard Deviation	Posterior Mode	Posterior Mean	Empirical Mean
10	900	.0111	.003	.0053	.0056	.0042
5	900	.0055	.0024	.0042	.0045	.0042
3	900	.0033	.0019	.0037	.0040	.0042
2	900	.002	.0016	.0035	.0038	.0042
1	900	.0011	.0011	.0033	.0036	.0042
10	1200	.008	.0026	.0050	.0052	.0042
5	1200	.00416	.0018	.0039	.0042	.0042
3	1200	.0025	.0014	.0035	.0037	.0042
1	1200	.0008	.0008	.0031	.0033	.0042

analysis of stay times in the system. Table 1 uses one example of a set of homogeneous data taken from an actual situation where people arriving to the system where counted using a video surveillance system [1]. Table 2 is a different set of data with homogeneous probabilistic behaviour. The data were studied separately then combined to provide a measure for the inter-arrival process.

For the stay times, we report an example for the purpose of illustration. The data is shown in Table 3. The number of people is counted for each cycle along with the total time of stay. In this example, we looked at 33 cycles.

Using only the data where the number of people in a cycle is 1, we conducted the estimation for the actual stay times. Figure 4 shows the statistical analysis of that data. Using a typical classical fit program, we found the mean to be 71.84 and the variance 989.16 for a Gamma distribution. The Gamma distribution had estimated shape parameter of 5.21 (1.64 standard error) and scale parameter 13.76 (4.54 standard error). In another analysis, an

Table 3 Total Time of Stay per Cycle

Cycle No.	Number of People	Total Time of Stay	Cycle No.	Number of People	Total Time of Stay
1	5	456	18	1	74
2	3	652	19	4	534
3	1	86	20	1	82
4	6	928	21	1	146
5	2	188	22	1	132
6	1	40	23	11	1096
7	2	212	24	8	476
8	6	2264	25	1	48
9	1	40	26	1	40
10	5	1035	27	5	394
11	1	63	28	11	1468
12	1	62	29	1	100
13	1	52	30	1	40
14	1	130	31	1	40
15	1	44	32	5	552
16	1	92	33	1	54
17	2	484			

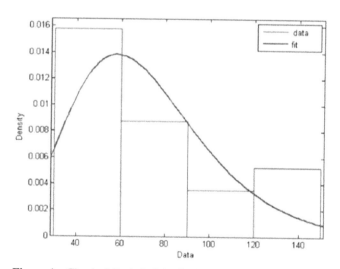

Figure 4 Classical Statistical Analysis of 1 person per Cycle Data

approximation to the actual stay times is obtained by dividing the total stay time in a cycle by the number of people in that cycle Y_i/n_i. Figure 5 shows the statistical analysis of that data. The distribution was found to be a Gamma with

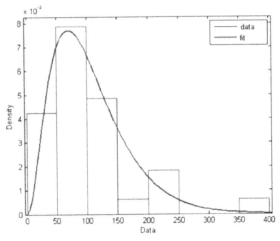

Figure 5   Classical Statistical Analysis of Averaged Stay Times Yi/ni.

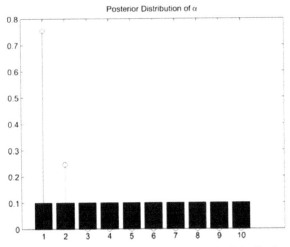

Figure 6   Uniform Prior Distribution and Posterior Distribution of α

estimated shape parameter of 2.98 (0.69 standard error) and scale parameter 35.3 (8.9 standard error). The distribution had a mean of 105.14 and variance 3712.

Using the full Bayesian analysis described in this report, Figure 6 shows the posterior distribution of α using all the data. It is customary in such a Bayesian methodology to use a uniform distribution if no prior knowledge exists about the parameter of interest, namely α in this case. This means that a-priori, all values of α are discrete and considered equally likely before the observation of the data. In our example, once the data has been accumulated and included in the Bayesian analysis, the first two values of α emerged as the most likely values, as shown in Figure 6

Figure 7 shows the prior distribution of λ and the two first posterior distributions obtained using 1 data point and two data points (more peaked distribution). As more and more data is introduced into the analysis, the posterior distribution becomes more concentrated around the posterior estimate

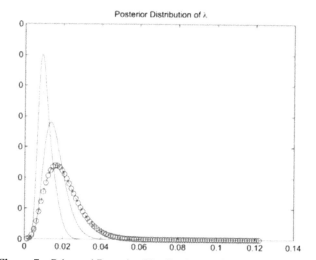

Figure 7 Prior and Posterior Distributions of λ with 1 and 2 Points

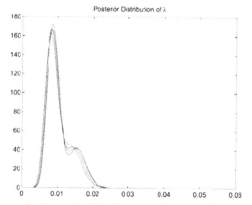

Figure 8 Posterior Distributions of λ with 8 data Points

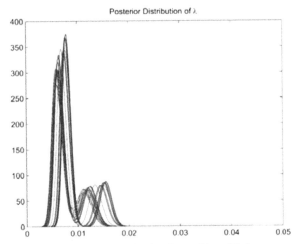

Figure 9   Posterior Distributions of λ at All Stages

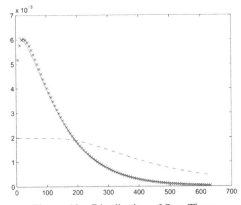

Figure 10   Distribution of Stay Times

of λ. This is typical of the behaviour of a posterior distribution. The Bayesian probabilistic updating algorithm provides a mean to refine an estimate within a probability interval as more and more data is gathered. At times, the data will also point to the possibility of bi-modality. This is often due to the mixing of two populations in the data, as was observed in our example when we introduced 8 data points. Figure 8 illustrates the posterior distribution with the first 8 data points, that is $p\,(\lambda|y_1), \ldots, P\,(\lambda|y_1, \ldots, y_8)$.

Finally, Figure 9 shows the posterior distributions at all the stages.

The interesting observation is the clear bimodality of the posterior distribution. This reflects the two types of people who enter the system, those who come in for a short period of time and those who stay longer. This corroborated our prior knowledge of the location we observed. Figure 10 shows the prior and posterior distribution of the stay time, after averaging over the parameters α and λ. The dashed plot is the prior distribution. The straight line is the distribution of $S_{i,j}$ averaged over α and λ. The x marked plot is a numerical approximation to the posterior distribution.

6 Conclusion

We present a Bayesian updating methodology for the parameters of the gamma distribution. We apply it to the study of the data of stay times in a system. Statistics for the stay times are deducted from the stochastic process of the number of incoming people. A gamma probability model is assumed for the stay times with unknown parameters. The Bayesian methodology is applied to update knowledge on these parameters probabilistically using the existing statistical information. Based on collected information on the number in the system, stay times are estimated in a probability model. The method is useful in the development of technology to estimate traffic in a system.

References

[1] S. Challa, K. Aboura, K. Ravikanth, S. Deshpande, 'Estimating the number of people in buildings using visual information', Information, Decision and Control, pp. 124–129 (2007).

[2] D. L. Snyder, M. I. Miller, 'Random Point Processes in Time and Space', Springer-Verlag, NJ (1991).

[3] D. V. Lindley, 'Reconciliation of probability distributions', Operations Research, pp. 866–880 (1983).

[4] K. Aboura, J. I. Agbinya, 'Adaptive maintenance optimization using initial reliability estimates', Journal of Green Engineering, pp. 121 (2013).

Biographies

Khalid Aboura teaches quantitative methods at the College of Business Administration, University of Dammam, Saudi Arabia. Khalid Aboura spent several years involved in academic research at the George Washington University, Washington D.C, U.S.A, where he completed the Master of Science and the Doctor of Science degrees in Operations Research. Dr Aboura has extensive experience in Stochastic Modelling, Operations Research, Simulation, Maintenance Optimization and Mathematical Optimization. He worked as a Research Scientist at the Commonwealth Scientific and Industrial Research Organization of Australia and conducted research at the School of Civil and Environmental Engineering and the School of Computing and Communication, University of Technology Sydney, Australia. Khalid Aboura was a Scientist at the Kuang-Chi Institute of Advanced Technology of Shenzhen, China.

Johnson I. Agbinya is an Associate Professor in the department of electronic engineering at La Trobe University, Melbourne, Australia. He is Honorary Professor at the University of Witwatersrand, South Africa, Extraordinary Professor at the University of the Western Cape, Cape Town and the Tshwane University of Technology, Pretoria, South Africa. Prior to joining La Trobe, he was Senior Research Scientist at CSIRO Telecommunications and Industrial Physics, Principal Research Engineering at Vodafone Australia and Senior Lecturer at University of Technology Sydney, Australia. His research activities cover remote sensing, Internet of things, bio-monitoring systems, wireless power transfer, mobile communications and biometrics systems. He has authored several technical books in telecommunications. He published more than 250 peer-reviewed research publications in international Journals and conference papers. He served as expert on several international grants reviews and was a rated researcher by the South African National Research Fund.

Comparative Analysis of Anisotropic Diffusion and Non Local Means on Ultrasound Images

Rishu Gupta[1], I. Elamvazuthi[1*], Ibrahima Faye[2], P. Vasant[2]
and J George[3]

[1]Department of Electrical and Electronic Engineering
[2]Department of Fundamental and Applied Sciences
Universiti Teknologi PETRONAS Perak, Malaysia
[3]Research Imaging Centre, University of Malaya, Malaysia
[1*]E-mail: irraivan_elamvazuthi@petronas.com.my (Corresponding Author)

Received 28 October 2012; Accepted 15 March 2013
Publication 23 January 2014

Abstract

Applications of Ultrasound in examining physical health of patient are continuously expanding over last few decades. Ultrasound imaging has assisted doctors in the diagnosis of medical health of patient in obstetrics, cardiology, gynaecology, musculoskeletal, urology and others. It has proved to be very promising and considerably advantageous over other imaging methods because of its features like non-invasive nature, use of non-ionizing radiation, portability, rapidity in performance, ease and cheap availability, real time imaging capability, high patient acceptability. Despite many advantages, it suffers from Speckle noise which results in affecting image resolution and contrast making an adverse impact on the diagnostic capability of the imaging modality. In this paper, we attempt to remove speckle noise from musculoskeletal ultrasound image for shoulder application using Anisotropic Diffusion and Non-Local Means (NLM) method. The quantitative analysis of the result is done using measurement indexes such as peak signal to noise ratio (PSNR), signal to noise ratio (SNR) and root mean square error (RMSE), whereas, visual inspection is carried out for qualitative analysis.

Journal of Machine to Machine Communications, Vol. 1 , 73–90.
doi: 10.13052/jmmc2246-137X.115

Keywords: Ultrasound Imaging, Speckle Noise, Anisotropic Diffusion, Non-Local Means, Quantitative Analysis.

1 Introduction

Ultrasound imaging has assisted radiologist and physicians in observing anatomy of human body since past few decades. Ultrasound was developed during the Second World War and since then is assisting for various applications on water, air and earth in various forms. Ultrasound as a medical imaging modality evolved about half a century ago and considered to be great invention because of its unique and advantageous features over all other existing medical imaging modalities. It offers several superior features such as use of non-ionizing radiation, real time imaging, portability, high patient acceptability, ease of use and is economical. Despite many advantages, ultrasound images contain noise known as 'Speckle' which corrupts the image resolution and contrast making interpretation of physical structure lying underneath, extremely difficult for physicians. Various phenomena like image acquisition and imperfections in machine design contribute to the existing resolution and contrast in ultrasound images. Despeckling of ultrasound images is considered to be of utmost important to make it a viable and suitable option for use as imaging modality. As a result, several efforts have been made by research community to despeckle the image as well as to improve the system design to help diagnostics.

Speckle noise in ultrasound image is introduced because of the coherence nature of imaging modality, the interference of back scattered signals from each resolution and sub-resolution cell towards receiver or transducer. The backscattered waves undergo constructive or destructive interference in a random manner spoiling image in a random granular pattern, termed as Speckle. Due to the acquisition process the speckle noise in ultrasound is multiplicative in nature which is directly proportional to grey level in any area and is statistically independent from signal.

The statistics of the speckle noise was first proposed by Goodman [1], where he modeled the statistical representation of speckle formation in laser. With study about statistical features in ultrasound imaging it was concluded that speckle in ultrasound images is multiplicative in nature following a different statistical distribution [22], [23], [24], [25] naming Rayleigh, Rician, Nakagami, K- distribution depending on the scatterer density and scatterer size of the tissue being imaged.

Noise in ultrasound images introduced has both multiplicative as well as additive nature making it more difficult to separate signals and denoise image. The multiplicative noise introduced due to the coherency feature whereas additive noise comes from the system properties. In the past few decades, efforts have been made to post process ultrasound images taken from system and remove noise from ultrasound images for better interpretation and diagnosis.

In this paper we have discussed in detail comparison about two despeckling method 1) Anisotropic diffusion method 2) Non Local Means Method. Anisotropic Diffusion Method [13], [14], [15] is successfully used to remove speckle noise from ultrasound images, whereas nonlocal means method [16] initially used for additive noise is recently introduced to despeckle ultrasound images [17]. The qualitative and quantitative comparison for two methods is also provided.

2 Speckle Noise

2.1 Model for Speckle Noise

The generalized model for speckle noise as proposed in [2] is given by:

$$g(n, m) = f(n, m) \text{ x } u(n, m) + \xi(n, m) \tag{1}$$

where, g, f, u and ξ are observed envelope image, original image, multiplicative noise and additive components respectively. The model has been successfully used both in Synthetic Aperture Radar as well as Ultrasound Imaging.

In the latter case, the model can be further simplified by disregarding the additive noise term. The image obtained without the application of system processing techniques, changes the resultant image model to:

$$g(n, m) = f(n, m) \text{ x } u(n, m) \tag{2}$$

An alternative model for speckle noise has been proposed in [3] wherein it is considered only as the additive noise, the amplitude of which is proportional to the square of the true image. The model is given by:

$$g(x) = x + u\sqrt{x} \tag{3}$$

where, $g(x)$ is the observed signal, x is the original signal without noise, u is the noise which is dependent of the observed signal.

The paper focuses on the B-Scan images taken from ultrasound machine focusing musculoskeletal application for shoulder images. In section (II), a brief discussion on the speckle reduction methods is provided combined with detail discussion for anisotropic diffusion and non-local means method, followed by, the description of materials and methods in section (III), discussion and conclusion about the findings are shown and summarized in section (IV) and section (V) respectively.

2.2 Speckle Reduction Methods

Speckle degrades information in the image and that too, to the extent, that it becomes difficult for medical practitioner to form a comprehensive understanding of tissues or bones for proper diagnosis.

Several researchers have studied this phenomena and attempted to improve the quality of images using different denoising methods like, mean filter, median filter, adaptive mean filters such as Kuan [4] and Lee [5], both filter have same characteristic and operate by computing the different combination of coefficient of variation of noised image and coefficient of variation of noise, in general operates on center pixel intensity and average pixel intensity of filter window, where filter forms a balance between averaging image in homogeneous region and applying identity filter elsewhere.

Frost [6], this filter also works on principle of averaging homogeneous region and all pass filter in other non-homogeneous sections. The difference lies in filter window which in this case is an exponentially shaped kernel that can form average filter or all pass filter on an adaptive basis. The response of the filter varies again with the coefficient of variation of noise and coefficient of variation of noised image.

Wavelet based speckle reduction methods are classified into Homomorphic and Non-Homomorphic wavelet filtering. In homomorphic method [7], [8], [9] multiplicative noise in ultrasound image is converted into additive noise by taking logarithmic transform in first step, followed by wavelet decomposition and modification of wavelet coefficients in second step, then inverse wavelet and exponential transform is performed to get back the reconstructed image. Whereas in non-homomorphic method [20], [21] it is considered that converting the multiplicative noise to additive noise results in loss of many statistical properties of the image and inappropriate performance of despeckling method [18], therefore in this method despeckling is performed without taking into account any noise model and the method can adapt itself to many different noise models present in medical images. In [12], we have analyzed ultrasound images using wavelets decomposition up to second level and performed qualitative as well as quantitative analysis using different measurement indexes.

Total variation denoising method is used to remove noise from an image model by solving minimization problem. The noise model for ultrasound image is formed and based on noise model regularization term and data fidelity terms are defined so as to despeckle the image. In [10], [11] the minimization problem is solved using novel Split- Bregman algorithm based on Zhang-Burger-Bresson-Osher's non local graph method.

The detailed discussion for anisotropic diffusion and Nonlocal means method is given in following section.

2.3 Anisotropic Diffusion Filtering

Anisotropic Diffusion filtering is a nonlinear filtering method coined by Perona and Malik [13] in the year 1990 for removing the additive Gaussian noise from images. In this method, partial differential equation (PDE) has been introduced for smoothing the image keeping track of homogeneous region and region containing edges in images. The nonlinear PDE for smoothing image introduced by Perona and Malik is:

$$\begin{cases} \frac{\partial I}{\partial t} = \partial iv[c(|\nabla I|).\nabla I] \\ I(t=0) = I_0 \end{cases} \tag{4}$$

where ∂iv is the divergence operator, $|\nabla I|$ is the gradient magnitude of the image I, $c(|\nabla I|)$ is the diffusivity function or the diffusion coefficient, I_0 is the original image. Perona and Malik [13] suggested two different functions for diffusion coefficients:

$$c(x) = \frac{1}{1 + (x/k)^2} \tag{5}$$

or,

$$c(x) = \exp[-(x/k)^2] \tag{6}$$

where k (set either manually or using noise estimator) is the parameter controlling the level of diffusion between edges and homogeneous region in images, optimum value should be chosen to avoid over or under smoothing. If $x >> k$, then $c(x)0$, we use all-pass filter, where as if $x << k$, then $c(x)1$, we use isotropic diffusion (Gaussian filtering). It is also suggested by [14], that anisotropic diffusion discussed above works well for additive noise where as in the presence of speckle noise, instead of delineating the image enhances speckle.

In [15], Yu and Acton proposed new anisotropic diffusion model to smooth speckle images where output image $I(x, y; t)$ is computed using the following differential equation:

$$\frac{\partial I(x, y; t)}{\partial t} = div[c(q)\nabla I(x, y; t)] \tag{7}$$

$$I(x, y; 0) = I_0(x, y), \tag{8}$$

$$\partial I(x, y; t)/\partial n)|_{\partial \Omega} = 0 \tag{9}$$

where, t represents diffusion time, $\partial \Omega$ denotes borders of Ω, n is outer normal to $\partial \Omega$, their diffusion coefficient $c(q)$ is written as

$$c(q) = \frac{1}{1 + [q^2(x, y; t) - q_0^2(t)]/[q_0^2(t)(1 + q_0^2(t))]} \tag{10}$$

or,

$$c(q) = \exp\{-[q^2(x, y; t) - q_0^2(t)]/[q_0^2(t)(1 + q_0^2(t))]\} \tag{11}$$

where $q(x, y ; t)$ is the instantaneous coefficient of variation at position (x, y) and is given by:

$$q(x, y; t) = \sqrt{\frac{(1/2)(\nabla I/I)^2 - (1/4^2)(\nabla^2 I/I)^2)}{[1 + (1/4)(\nabla^2 I/I)]^2}} \tag{12}$$

The function $q(x,y;t)$, instantaneous coefficient of variation combines normalized Laplacian operator and normalized gradient operator to detect edges from the speckled images. At the edges, Laplacian term shows zero crossing whereas gradient gives higher values allowing detection of edges in bright as well as dark regions. The function takes higher values at edges or high contrast regions, whereas, low values in homogeneous regions.

2.4 Non Local Means (NLM)

The local smoothing or frequency domain filter takes advantage of regularized geometrical configuration in images, while neglecting details and fine structures preservation. Due to regularity assumption, fine structures like edges and other details are smoothed in images leading to information loss. In [16], Antoni Buades introduced non-local means (NLM) algorithm which takes advantage of redundant features in natural image. Redundant features refer to the similar patches in image which very often are present in natural images. Mathematically, NLM algorithm is described as, for a given noisy image $u(x)$, denoised value for pixel can be written as,

$$u'(x) = \frac{1}{C(x)} \sum_{x' \in \Omega_s} w(x, x') f(x') \tag{13}$$

Where $C(x)$ is the normalizing factor and $w(x, x')$ is the weight determined by the similarity level of different patches around x, x'. The weight $w(x,x')$ is computed using

$$w(x, x') = \exp(-\|P(x) - P(x')\|_{2,a}^2 / h^2) \tag{14}$$

where, $P(x)$ denotes the square patch centered around pixel x, h is filtering parameter and $\|\bullet\|_{2,a}$ is a Gaussian weighted Euclidean distance whose value is determined by amount of similarity between patches from different neighborhood. Higher the similarity, more will be $w(x,x')$.

According to [17], applying NLM directly on the speckle noise or ultrasound images will not yield better results because of the multiplicative nature of speckle contained in the images. Speckle in ultrasound images is different from additive noise, therefore using square patches yields ringing artifacts, noise named halo. In eq(15), $S(\tau)$ gives definition to different shaped patches like disks and pies which could effectively deal with speckle noise compared to square patches which were introduced initially. The author in [18] suggested Gaussianizing speckle noise in ultrasound images, before using non local means-multi shape normal patches (NLM-MSPA). The measurement between the pixels in Eq. (14) is reformulated as:

$$w_s^2(x, x') = \sum_{r \in \Omega} S(\tau)(f(x + \tau) - f(x' + \tau))^2 \tag{15}$$

where, S gives definition to the patch considered which could be, for instance it can be written as:

$$S(\tau) = \begin{cases} 1, if \|\tau\|_\infty \le \frac{p-1}{2} \\ 0, otherwise \end{cases} \tag{16}$$

This makes patch as standard square patch of size of pxp and the distance being measured as Euclidean norm.

Two issues prevail when it comes to taking into consideration arbitrary patches for distance measure between pixels, first is to choose the shape among the existing ones and second is to combine the estimator so as to give a combined result from every estimate. To solve first problem Stein proposition [26] unbiased estimate for risk estimation for pixel x is considered and shape based norm defined in Eq(15) has been modified to following derivative expression

$$\frac{\partial w(x, x')}{\partial \varepsilon(x')} = \frac{\left(S(0) \left[f(x) - f(x') \right] + S(x - x') \left[f(x) - f(2x - x') \right] \right)}{h^2}$$

(17)

where S defines shape of Kth estimator. The second concern is dealt with statistical method of exponentially weighted aggregation (EWA) [27], where several estimators are combined using weighted average, giving higher weights to estimators with low risks.

3 Material and Methods

Ultrasound images were taken from a healthy shoulder of ten different volunteers after all subjects gave their informed consent before participation and the procedures were approved by the local ethics committee. The experimental setup was designed so as to acquire best possible image with maximum comfort of subject. Figure 1 provides the setup configuration.

Despeckling was done using anisotropic diffusion and nonlocal means algorithm. For rigorous study, 70 different images from ten subjects were involved and quantitative parameters were recorded for each image to establish concrete foundation for results. In anisotropic method, the function used for denoising is taken from Eq. 5, and number of iterations was set to 10. In

Figure 1   Experimental setup for Data Acquisition

the NLM method, square patches were taken and weight is calculated using Euclidean distance formula introduced in Eq. (14).

In order to quantitatively compare the performance, three different measurement indexes such as peak signal to noise ratio (PSNR), root mean square error (RMSE) and signal to noise ratio (SNR) [19] have been used. The PSNR is calculated using:

$$PSNR = 10.\log_{10}\left(\frac{M^2}{E\phi_i^j(x)[x - \hat{x}]^2}\right) \qquad (18)$$

where, M is the peak pixel amplitude, x represents pixel amplitude in the original image, and $\hat{x}$ is the corresponding pixel in reconstructed image. The greater the PSNR better is the result.

MSE is measure of error between the filtered or denoised image and original image. The greater the value poorer is the result.

$$RMSE = \frac{1}{MN}\sum_{x=1}^{M}\sum_{y=1}^{N}[I(x,y) - \hat{I}(x,y)]^2 \qquad (19)$$

where, $I(x, y)$ is the original image and $\hat{I}(x, y)$ is the denoised image and MN is the size of image.

Third measure indexes introduced for quantitative analysis was signal to noise ratio (SNR). SNR value is calculated using the formulae

$$SNR = \frac{\mu}{\sigma} \qquad (20)$$

where, μ is the mean calculated over the resultant image taken into consideration and σ is the standard deviation of the image.

4 Results and Discussion

Figure 2 shows the original image in first column, followed by same image processed using anisotropic diffusion in second column and non-local means algorithm in the third column.

From Figure 2, it can be seen that, anisotropic diffusion gives the smoother image with 10 iterations which on increasing number of iteration becomes blurry and loose important details for diagnosis, on the other hand nonlocal means technique preserves the important details in image at same time maintaining sharpness in image resulting in better contrast. In image 1,

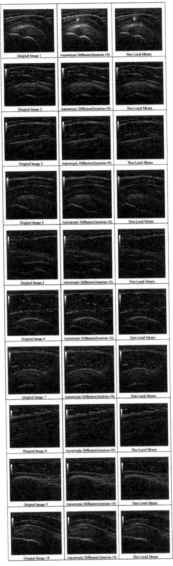

Figure 2 Results obtained with Anisotropic Diffusion and Non local means method on shoulder ultrasound images.

having a closer look provides necessary interpretation about two techniques discussed. Yellow arrow, indicated in both processed image shows blurriness at edges in anisotropic diffusion method whereas sharpness using nonlocal means method, red arrow shown in homogeneous region gives details about the smooth structure inside the healthy tendon of a subject which is more prominent and sharp after using nonlocal means method. The visual results from all images considered are in compliance with [17]. The blurriness in the image processed by anisotropic diffusion increases as we increase the number of iterations, which makes homogeneous regions smooth at the same time hides important details. NLM algorithm takes into account redundancy in the patches and iterations are not required for smoothing, therefore the smoothing of homogeneous regions and sharpness in edges, the better patches and weight calculation leads to more optimum results.

The PSNR, SNR and RMSE values obtained for the ultrasound image despeckling using the anisotropic diffusion and nonlocal means algorithm are shown in Table 1.

It can be seen that the average PSNR value for non-local means method is 35.2844 whereas for anisotropic diffusion it is computed as 32.8985 which is well below NLM method. Also, the computed average RMSE value for NLM method is 0.1721 whereas that for anisotropic diffusion method is 0.2279. The SNR value computed for every image shows significant increase in the signal to noise ratio in case of NLM method as compared to anisotropic Diffusion. Taking into account the statistical relevance provided by all three measurement metric of anisotropic diffusion and its compliance with visual interpretation, it is clear that NLM algorithm performs well for ultrasound images as compared to anisotropic diffusion method.

Table 1   Comparison Results for despeckling

Methods	ANISOTROPIC DIFF. (ITER-10,FN-1)			NON LOCAL MEANS (NLM)		
Images	PSNR	SNR	RMSE	PSNR	SNR	RMSE
Image 1	30.2541	1.3915	0.0303	34.0893	1.58	0.01951
Image 2	31.7288	1.2468	0.0259	34.5822	1.29	0.01865
Image 3	33.3226	1.32	0.0215	35.4642	1.346	0.01685
Image 4	33.7039	1.2841	0.0206	35.6655	1.34	0.01647
Image 5	33.9778	1.2606	0.0200	35.8742	1.3125	0.01607
Image 6	33.1249	1.44	0.0220	35.4796	1.4933	0.01682
Image 7	33.7199	1.31	0.0206	35.6103	1.325	0.01657
Image 8	32.9876	1.2887	0.0224	35.1445	1.327	0.01748
Image 9	33.3162	1.203	0.0215	35.6129	1.2284	0.01657
Image 10	32.8491	1.0644	0.0227	35.3206	1.11	0.01713

The trend for the variation in values of PSNR, SNR and RMSE is also shown in graphical form in Figure 3 for better understanding.

In Figure 3, it can be clearly seen that values for PSNR and SNR is greater for every image in NLM method than for anisotropic diffusion method, and value for root mean square error is smaller for every image for NLM method

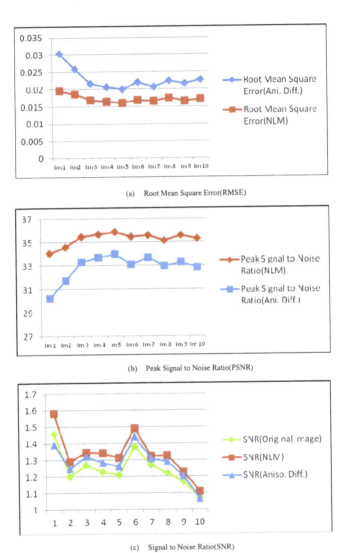

(a)   Root Mean Square Error(RMSE)

(b)   Peak Signal to Noise Ratio(PSNR)

(c)   Signal to Noise Ratio(SNR)

Figure 3   Graphical representation of Measurement indexes.

compared to that for anisotropic diffusion method. In general, it can be deduced that the results obtained are in compliance with the analytical discussion as well as in accordance with [17]. Further work in direction of qualitative analysis will be carried out in future taking into account necessity and needs of medical practitioners.

5 Conclusion

The findings from the study on anisotropic diffusion and NLM methods, it can be concluded that NLM method is superior and yields better results for ultrasound image despeckling both qualitatively and quantitatively. Further work involving radiologist and clinicians will be carried out to validate results qualitatively and focus on enhancing the NLM algorithm with various variants so as to provide more accurate results for better diagnosis and understanding.

6 Acknowledgement

The authors would like to thank UTP for their assistance and Ministry of Science, Technology and Innovation for sponsoring the project under the E-Science Grant Scheme: E-Science-03-02-02-SF0109.

References

[1] J. W. Goodman, "Statistical properties of laser speckle patterns," in Laser Speckle and Related Phenomena. J. C. Dainty, Ed. Berlin: Springer-Verlag, pp. 9–77 (1977).

[2] A. K. Jain, "Fundamental of Digital Image Processing," Englewood Cliffs, NJ: Prentice-Hall (1989).

[3] T. Loupas, W. N. Mcdicken, and P. L. Allan, "An adaptive weighted median filter for speckle suppression in medicalultrasonic images," IEEE Trans. Circuits Syst., vol. **36**, pp. 129–135, Jan. (1989).

[4] D. T. Kuan, A. A. Sawchuk, T. C. Strand, and P. Chavel. "Adaptive restoration of images with speckle." IEEE Trans. ASSP. vol.**35**, no.3, pp. 373–383, March (1987).

[5] J. S. Lee, "Redefined filtering of image noise using local statistics." Journal of Computer Graphic and Image Processing. vol.**15**, pp. 380–389, March (1987).

[6] V. S. Frost, J. A. Stiles, K. S. Shanmugam and J. C. Holtzman, "A model for radar images and its application for adaptive digital filtering of multiplicative noise." IEEE Transactions on pattern analysis and machine intelligence, vol.**4**, No.2, pp. 157–165 (1982).

[7] R. K. Mukkavilli, J. S. Sahambi, Bora, P. K. "Modified homomorphic wavelet based despeckling of medical Ultrasound images." IEEE CCEC, 000887–000890, July 2008.

[8] A. Pizurica, W. Philips, and I. Lemahieu, et al, "A versatile wavelet domain noise filtration technique for medical imaging," IEEE Trans. On Med. Imaging, vol. **22**, no.3, pp. 323–331, Mar. (2003).

[9] D. Lai, N. Rao, C. H. Kuo, S. Bhatt, V. Dogra, "An ultrasound image despeckling method using independent component analysis," IEEE ISBI, pp. 658–661, Aug. (2009).

[10] L. Wang, L. Xiao, L. Huang, Z. Wei "Nonlocal total variation based speckle noise removal method for ultrasound image" 978-1-4244-9306-7, International congress on Image and Signal Processing. (2011).

[11] M. N. Kohan, H. Behnam "Denoising Medical Images using Calculus of Variations" J Med Signals Sens., **1**(3): 184–190, Jul-Sep; (2011).

[12] R. Gupta, I. Elamvazuthi and P. Vasant. "Musculoskeletal Ultrasound image Denoising using Daubechies Wavelets", Global Conference on Power Control and Optimization, PCO Global August (2012).

[13] P. Perona, J. Malik. "Scale space and edge detection using anisotropic diffusion." IEEE Transaction Pattern Analysis Machine Intell. **12** 629–639 (1990).

[14] X. Zhi, T. Wang "An Anisotropic Diffusion Filter for Ultrasonic Speckle Reduction". Fifth IEEE international conference on Visual information engg., VIE., 327–330 (2008).

[15] Y. Yu and S. Acton, "Speckle reducing anisotropic diffusion", IEEE Trans. Image Process., vol.**11**, pp. 1260–1270, Nov. (2002).

[16] A. Buades, B. Coll, and J. M. Morel, "A non-local algorithm for image denoising," in Computer Vision and Pattern Recognition, 2005. CVPR 2005. IEEE Computer Society Conference on, vol.**2**, pp. 60–65 (2005).

[17] W. Chen, M. Ding, Y. Miao, L. Luo, "Ultrasound image denoising with multi-shape patches aggregation based non-local means," IEEE ICBMI, pp. 14–17, Dec. (2011).

[18] O. V. Michailovich, and A. Tannenbaum, "Despeckling of medical ultrasound images," IEEE Trans. on Ultrason., Ferroelect., Freq. Contr., vol. **53**, no. 1, pp. 64–78, Jan. (2006).

[19] S. Kalaivani Narayanan and R. S. D. Wahidabanu "A View on Despeckling in Ultrasound Imaging", International Journal of Signal Processing, Image Processing and Pattern Recognition Vol. **2**, No.3, September (2009).

[20] S. Yan, J. Yuan, M. Liu, and C. Hou, "Speckle Noise Reduction of Ultrasound Images Based on an Undecimated Wavelet Packet Transform Domain Nonhomomorphic Filtering," IEEE 978-1-4244-4134, Oct (2009).

[21] S. Gupta, RC Chauhan, SC. Saxena, "Robust non-homomorphic approach for speckle reduction in medical ultrasound images", Med Biol Eng Comput. **43**(2):189–95, Mar. (2005).

[22] V. Dutt and J. F. Greenleaf, Ultrasound echo envelope analysis using a homodyned K distribution signal model," Ultrason. Imag., vol. **16**, pp. 265–287 (1994).

[23] R. C. Molthen, P. M. Shankar, and J. M. Reid, Characterization of ultrasonic B-scans using non-Rayleigh statistics," Ultrasound Med. Biol., vol. **21**, no. 2, pp. 161–170 (1995).

[24] V. M. Narayanan, P. M. Shankar, and J. M. Reid, Non-Rayleigh statistics of ultrasonic back scattered signals," IEEE Trans. Ultrason., Ferroelect., Freq. Contr., vol. **41**, no. 6, pp. 845–852, Nov. (1994).

[25] P. M. Shankar, V. A. Dumane, J. M. Reid, V. Genis, F. Forsberg, C. W. Piccoli, and B. B. Goldberg, Classication of Ultrasonic B-Mode Images of Breast Masses Using Nakagami Distribution IEEE Transactions on ultrasonics, ferroelectrics, and frequency control, vol. **48**, no. 2, March (2001).

[26] D. V. D. Ville and M. Kocher, "SURE-Based Non-Local Means," Signal Processing Letters, IEEE, vol. **16**, pp. 973–976 (2009).

[27] A. Dalalyan and A. B. Tsybakov, "Aggregation by exponential weighting, sharp PAC-Bayesian bounds and sparsity," Mach. Learn., vol. **72**, pp. 39–61 (2008).

Biographies

Rishu Gupta was born in Gonda, UP, India, in 1984. He received his B. Tech degree in Electronics and Communications Engineering from Bundelkhand Institute of Engineering and Technology in 2009 and M.S degree in Visual Contents from Dongseo University, South Korea. Currently, he is working toward the PhD Degree from UniversitiTeknologiPetronas, Malaysia. He is active IEEE student member since 2012. His research interests include medical ultrasound image processing and Visual Contents.

I. Elamvazuthi obtained his PhD from the Department of Automatic Control and Systems Engineering, University of Sheffield, UK in 2002. Dr. I. Elamvazuthi is an Associate Professor at the Department of Electrical and Electronic Engineering of UniversitiTeknologi PETRONAS, Malaysia. He is a member of IEEE, IFAC, AEE, AAIA and BEM. His research interests include Control and Systems Engineering with focus on energy (power systems), robotics and bio-medical applications.

Ibrahima Faye is an Associate Professor at UniversitiTeknologi PETRONAS, Tronoh, Malaysia. He is attached to the Department of Fundamental and Applied Sciences and the Centre for Intelligent Signal and Imaging Research. He received a BSc, MSc and PhD in Mathematics from University of Toulouse and a MS in Engineering of Medical and Bio-technological data from EcoleCentrale Paris. His research interests include Engineering Mathematics, Signal and Image Processing, Pattern Recognition, and Dynamical Systems.

P. Vasant is a senior lecturer in the Department of Fundamental and Applied Sciences, UniversitiTeknologi PETRONAS (UTP), Malaysia. His research interests are Hybrid optimization, Soft computing and Computational intelligence.

Professor Dr JGeorge is Head of Musculoskeletal Radiology at the Department of Biomedical Imaging, University of Malaya Medical Centre in Kuala Lumpur, Malaysia. He has spoken at several International Conferences and runs annual Musculoskeletal MRI and ultrasound workshops in Penang Island. He has authored about 35 ISI cited publications and his research interests are in menisco-capsular tears of the knee, knee cartilage imaging and classification, MRI for age estimation used by FIFA and use of High Resolution of Ultrasound for arthritis screening.

Online Manuscript Submission

The link for submission is: www.riverpublishers.com/journal

Authors and reviewers can easily set up an account and log in to submit or review papers.

Submission formats for manuscripts: LaTeX, Word, WordPerfect, RTF, TXT.
Submission formats for figures: EPS, TIFF, GIF, JPEG, PPT and Postscript.

LaTeX

For submission in LaTeX, River Publishers has developed a River stylefile, which can be downloaded from http://riverpublishers.com/river publishers/authors.php

Guidelines for Manuscripts

Please use the Authors' Guidelines for the preparation of manuscripts, which can be downloaded from http://riverpublishers.com/river publishers/authors.php

In case of difficulties while submitting or other inquiries, please get in touch with us by clicking CONTACT on the journal's site or sending an e-mail to: info@riverpublishers.com